Ulrich Meve

The Genus Duvalia (Stapelieae)

Stem-Succulents between the Cape and Arabia

SpringerWienNewYork

Dr Ulrich Meve
Institut für Botanik, Westfälische Wilhelms-Universität,
Münster, Bundesrepublik Deutschland

Softcover reprint of the hardcover 1st edition 1997

Typesetting: Thomson Press (India) Ltd., New Delhi

Graphic design: Ecke Bonk

Cover illustration: *Duvalia polita* in habitat (Zimbabwe: Bulawayo; growing in association with *Ceropegia rendallii* N. E. Br.). Insert: *Duvalia polita*, flower (Zimbabwe: Bulawayo, Meve & al. 532, flowering in cultivation).

Printed on acid-free and chlorine free bleached paper

SPIN: 106 286 48

With 54 Figures and 4 coloured Plates

Library of Congress Cataloging-in-Publication Data

Meve, Ulrich, 1958–
The genus Duvalia (Stapelieae) : stem-succulents between the Cape and Arabia / Ulrich Meve.
p. cm. – (Plant systematics and evolution. Supplement, ISSN 0172-6668 ; 10)
Includes bibliographical references and index.
ISBN-13: 978-3-211-82984-4 e-ISBN-13: 978-3-7091-6854-7
DOI: 10.1007/978-3-7091-6854-7
1. Duvalia–Africa, Southern–Classification. 2. Duvalia–Arabian Peninsula–Classification. I. Title. II. Series: Plant systematics and evolution. Supplementum ; 10.
QK495.A815M48 1997
583′.93–dc21

Foreword

Dr ULRICH MEVE's book on *Duvalia*, one of the 30 genera comprising the group of stem-succulent *Asclepiadaceae*, is a piece of well-done research. The text is supported by excellent illustrations comprising full-page drawings and photographs making this monograph a very practicable contribution for both professional and amateur botanists. Dr MEVE's species delimitations are mainly based on morphological, karyological and biogeographical studies, and therefore easy to follow. His taxonomic and well-established biogeographic conclusions have been supported by field observations in southern and eastern Africa. Without doubt, this book represents one of the best and most intensive monographs ever published for the *Asclepiadaceae*.

The studies of Dr MEVE were initiated at my former chair of systematic botany in Kiel, the former work place of the renowned succulent plants specialist Dr H. JACOBSEN and the karyologist Prof. G. REESE. Dr MEVE has been involved with *Asclepiadaceae* research for more than ten years, and his work is characterized by a meticulousness typical of him. He has made himself a name amongst colleagues as well as plant lovers by his remarkable number of scientific and popular publications, which are not restricted to succulent *Asclepiadaceae*.

July, 1997

Prof. Dr Focke Albers
IOS Coordinator
Münster University, Germany

Preface

The stapeliad bibliographic history of the early 20th Century has been mostly influenced by N. E. Brown's treatment of the *Asclepiadaceae* in the "Flora Capensis" (1908) and the still very much used three volume treatise "The Stapelieae" of White & Sloane (1937). Since then, comprehensive treatments of the stapeliads in general are missing except for Lückhoff's (1952) work on southern African taxa. Taxonomic treatments of single genera followed such as those of L. C. Leach, who published a series of generic revisions (*Orbea, Orbeopsis, Tridentea, Stapeliea, Huernia*, 1978-1988). The late Leach was succeeded by P. V. Bruyns, whose outstanding revisions are based on intensive field work and are exceptionally well-illustrated (*Quaqua, Pectinaria, Echidnopsis, Lavrania, Hoodia, Tromotriche* etc., 1983 ff.). Unlike the previous revisions, this publication goes beyond the limits of morphological investigations since I endeavoured to shed light onto the overall biological behaviour of *Duvalia*. *Duvalia* served as a perfect candidate for studying the reproductive system of a genus, which has hitherto not been done in the *Asclepiadaceae*. Lucky circumstances aided the realization of crossing experiments and their interpretation:

1. Field studies and collection of *Duvalia* are comparatively easy, because most species are distributed in SW South Africa at many localities which are not too far apart and are easily accessible.
2. *Duvalia* performs well in cultivation and often flowers readily in the greenhouse, which facilitates determination as well as offering many artificial crossing opportunities.
3. Hand pollination can be done easily, because of the large pollinaria and nicely accessible guide rails.
4. *Duvalia* shows different levels of polyploidisation – unlike most other *Asclepiadaceae* – which are responsible for the development of some interesting reproductive barriers.

An increasing fascination in the enormous range of structures, colours and sizes of stapeliads (or *Asclepiadaceae* in general) can be observed in contrast to the decreasing fascination of "classical" succulents (e.g., *Cactaceae*). However, stapeliads are still unknown to many people. I hope my work will contribute to intensify the interest in this family and will serve as a stimulus for visiting and observing these plants in their beautiful and diverse habitats.

This book would not have become a reality without the support of many friends and colleagues: I am deeply indebted to Prof. F. Albers, Münster University, for supervision, maintenance and support. For technical assistance and pleasant cooperation in the laboratory, I would especially like to thank Mrs M. Frönd. The Directors and Curators of the following herbaria and institutions allowed access to their material: B, BM, BOL, FT, G, K, NBG, P, PRE, SRGH, WIND, ZSS. The National Botanical Institute (NBI), Pretoria, assisted with the identification of non-stapeliad specimens. Many friends and colleagues supported the work by contributing plant material, photographs, locality data etc. I would especially like to mention Dr G. BARAD, USA; M. B. BAYER, South Africa; A. DEBOER, The Netherlands; Dr W. BOSMA, The Netherlands; I. PEHLEMANN-BRASE, Namibia; Dr P. BRUYNS, South Africa; S. COLLENETTE, England; Dr U. DEIL, Germany; Dr U. EGGLI, Switzerland; Drs M. & M. GERBAULET-STRUCK, Germany; M. G. GILBERT, England; Dr H. E. K. HARTMANN, Germany; E. VAN JAARSVELD, South Africa; Prof. N. JÜRGENS, Germany; J. J. LAVRANOS, Portugal; F. NOLTEE, The Netherlands and D. PLOWES, Zimbabwe. Prof. S. LIEDE and P. S. MASINDE took care of polishing out the English manuscript. Finally, I would like to thank the editor and the referees for their various proposals to improve the manuscript.

July, 1997 *Ulrich Meve*

Contents

Introduction . . . 1
  Material and methods . . . 2
General part . . . 4
  Distribution and habitats . . . 4
  Morphology . . . 11
  Chromosomes and ploidy levels . . . 28
  Pollination . . . 33
  Breeding system . . . 33
  Flavonoids . . . 36
  Phylogenetic and biogeographic considerations . . . 37
Taxonomy . . . 45
  Taxonomic history . . . 45
  Taxonomic treatment . . . 45
    Key to the species (sect. *Duvalia*) . . . 47
    1. *Duvalia angustiloba* . . . 49
    2. *Duvalia caespitosa* . . . 52
    3. *Duvalia corderoyi* . . . 65
    4. *Duvalia elegans* . . . 70
    5. *Duvalia gracilis* . . . 75
    6. *Duvalia immaculata* . . . 78
    7. *Duvalia maculata* . . . 81
    8. *Duvalia modesta* . . . 84
    9. *Duvalia parviflora* . . . 87
    10. *Duvalia pillansii* . . . 90
    11. *Duvalia polita* . . . 92
    12. *Duvalia pubescens* . . . 99
    13. *Duvalia vestita* . . . 102
    Key to the species (sect. *Arabica*) . . . 107
    14. *Duvalia eilensis* . . . 108
    15. *Duvalia galgallensis* . . . 110
    16. *Duvalia sulcata* . . . 113
    17. *Duvalia velutina* . . . 121
References . . . 126
Index to descriptions of taxa accepted and synonyms . . . 131

Listed in Current Contents

Introduction

Of the 20 stapeliad genera described for southern Africa, the genus *Duvalia* Haw. shows many similarities to *Piaranthus* R. Br., which has been investigated in a parallel study (Meve 1994). The difficulty in separating the two genera based on sterile material is well known: strikingly similar characters are the plagiotropic growth forming flat clumps or mats, and the small ovoid-cylindrical stems with small triangular leaf rudiments associated with globoid stipular glands. Nevertheless, the mentioned and the present study suggest that the two genera belong to different lineages and obviously have developed in parallel due to ecological conditions in southern Africa. In flower, *Duvalia* can easily be separated from all stapeliad genera by the corolla tube being raised to form an annulus. This surrounds or is more or less covered by a coronal disc which closes at least the bottom of the corolla tube. The disc is formed by fused interstaminal and staminal corona parts of the stipitate gynostegium. The crown-like staminal corona, formed by two-parted staminal corona lobes, of which the inner part is horizontal and incumbent on the anthers, also marks unmisleadingly the genus as well as the broadly winged translator of the pollinaria.

Separation of the species is not easy, because there are only few good and non-quantitative characters. The shortage of characters and a high variability in the character states, is a principal problem in *Stapelieae* taxonomy at species level. Stuessy's (1990) thesis: "If... the phenetic groups... are similar, they need to be examined in more detail for their geographic, reproductive and ecological attributes. If they are sympatric, they will be good species if reproductively isolated", perfectly fits the species concept used in this study. It further implies that morphologically similar or even groups of hybrid origin may be treated as good species (cf. Löve 1951, Stace 1989), here for instance *D. elegans* and *D. vestita*, while phenetically quite divergent populations may be combined under one species, e.g., *D. caespitosa*.

These difficulties described above and the many allopolyploids and hybrids possibly involved do hamper a "cladistic analysis" considerably (cf. Bachmann 1995), which is therefore not presented here.

In the present study 17 species are recognized. One species is subdivided into subspecies and another into varieties. Stuessy's (1990) infraspecific concept is adopted: subspecies are more or less allopatric, possess several conspicuous morphological differences and rarely hybridize along contact zones. In contrast, varieties are largely allopatric with some overlap, possess one or two morphological differences and probably hybridize in overlap zones.

None of the *Duvalia* species is known to have played a major role in botanical history, nor is there a *Duvalia* species of economic importance. Horticulturally, *Duvalia* is of less interest even among stapeliad enthusiasts, as the plants are small-stemmed and require considerable cultivation efforts due to the fast growth of the relatively short-lived stems. Moreover, their medium-sized flowers usually occur singly and are not too showy. However, in contrast to many other stapeliads, they are easy to flower, and in details of stem and flower morphology a fascinating diversity can be observed. Distribution ranges from the Arabian Peninsula to the southwestern Cape, with a conspicuous gap in East Africa. Based on a sound morphological analysis and especially on the investigation of chromosome numbers of all taxa, an explanation of the puzzling distribution will be presented, allowing phylogenetic conclusions beyond those previously available for asclepiad genera.

Abstract

Based on analysis of vegetative and floral morphology, karyology and floral flavonoids, the genus *Duvalia* is revised. Seventeen species are recognized, one is described as new and three are reduced to infraspecific rank. All taxa are illustrated and distribution maps are provided. The species can be referred to two sections that match with the area of distribution: (1) sect. *Duvalia*, southern Africa, with *D. angustiloba, D. caespitosa, D. corderyoi, D. elegans, D. gracilis, D. immaculata, D. maculata, D. modesta, D. parviflora, D. pillansii, D. polita, D. pubescens* and *D. vestita*, and (2) sect. *Arabica*, NE Africa and SW Arabian Peninsula, with *D. eilensis, D. galgallensis, D. sulcata* and *D. velutina*. In sect. *Duvalia* there occur di-, tetra- and hexaploids, while in sect. *Arabica* only diploid taxa are found together with tetraploid cytotypes. Polyploidy is obviously of major significance for the speciation and distribution of the S African taxa. This is also corroborated by hybridisation experiments. Some of the diploid species are rare and have a narrow distribution, indicating their relict state. The small number of components of flavonoid aglycons found in sect. *Duvalia* is still more reduced in sect. *Arabica*. – The chromosome and biogeographical data suggest that the origin of *Duvalia* was in East Africa, perhaps in the Horn of Africa. The present distribution is obviously due to north- and southward migration. The sister genus of *Duvalia* is probably *Huernia*, while similarities with *Piaranthus* are obviously due to parallel evolution.

Material and methods

Material. The study is based on the investigation of ca. 250 specimens and 250 living plants. The living plants were cultivated by the author or are in cultivation at Münster Botanical Garden, Münster, Germany. The wild material originates predominantly from collecting trips of ALBERS, MEVE & KUSCH 1983, MEVE 1986, and MEVE & LIEDE 1988 to South Africa, Namibia and Zimbabwe, as well as from numerous botanists, collectors and institutions, in particular M.B. BAYER, Worcester, Dr P. BRUYNS, Cape Town, D. HARDY, Pretoria, J. LAVRANOS, Loule, I. PEHLEMANN-BRASE, Windhoek, and D. PLOWES, Mutare.

I had access to material from the following herbaria: B, BM, BOL, FT, G, K, NBG, P, PRE, SRGH, WIND and ZSS.

Methods. Morphology. Morphological studies were carried out with a Wild M3 stereo microscope, a Leitz microscope (Orthoplan), and scanning electron microscope (Leitz AMR-100 and Hitachi S53). The samples for the latter were critical point dried (CP-Dryer from Balzers Union), sputtered with gold and scanned at 10–20 kV.

C h r o m o s o m e c o u n t s . Chromosome numbers were established from adventitious root tip squash preparations. About 200 different collections cultivated in the local greenhouse were investigated. The root tips were prepared as follows: pretreatment in 0.002 M hydroxyquinoline for 4 h at 20 °C (TJIO & LEVAN 1950); fixation in Carnoys solution for 24 h at 20 °C; stained in carmine for 24 h at 60 °C (SNOW 1963). Vouchers for each taxon are marked by an asterisk (*) in "Specimens examined".

F l a v o n o i d i n v e s t i g a t i o n s . Acetone/methanol extracts of stems or flowers were hydrolysed in boiling HCl for 40 min. The aglycon fractions were gained by extracting the products of hydrolysis with ethanolacetate. The flavonoids were identified by chromatographic comparison on cellulose TLC-sheets against authentic samples of apigenin, isorhamnetin, kaempferol, luteolin, myricetin and quercetin. The solvent used was CAW (5:4). After treatment with 5% methanolic $AlCl_3$ the chromatograms were analysed under UV-light (366 nm).

L i t e r a t u r e c i t e d . Literature appearing in the "Taxonomic treatment" is not given in the reference list. Author abbreviations follow BRUMMIT & POWELL (1992). Distribution data for the S African taxa are given in relation to the "geographical quarter degree grid square system" of EDWARDS & LEISTNER (1971).

General part

Distribution and habitats

Distribution and phytochorological association. The distribution of the genus *Duvalia* is disjunct with one partial distribution area in southern Africa (sect. *Duvalia*) and another in northwestern Africa and southwestern Arabia (sect. *Arabica*, Fig. 1). The distribution gap starts in Somalia, spans nearly 3000 km and ends with the occurrence of *D. polita* (sect. *Duvalia*) in Malawi, Zambia, Angola, etc. This allopatric species is restricted to the "Sudano-Sambesian Region" (sensu WERGER 1978). All the other species of sect. *Duvalia* occur in the Cape Provinces of South Africa, from where few species radiate into the southern Namib or the Free State (Figs. 1, 2). One of the two centres of diversification is the Little Karoo of the Western Cape with 4 species /100 km^2 (Fig. 2). The Little Karoo is part of the "Succulent Karoo Region" (JÜRGENS 1991), which is known for its richness in endemic species, as well as for its extraordinary diversity. In the south this region merges into the "Cape Floristic Region" (=Capensis sensu WHITE 1983), where only two *Duvalia* species can be found. Both regions form the "Greater Cape Flora" *sensu* JÜRGENS (1991), which receives mostly winter rainfall. In the southern and eastern Cape Provinces, however, there is an increasing influence of summer rainfall along the transition into the "Nama-Karoo Region" (JÜRGENS 1991). Here, around 32° S 24° E, the occurrence of *Duvalia* increases considerably and reaches the highest concentration with 6 species/100 km^2 (Fig. 2).

Section *Arabica* occurs mainly around the lower Red Sea and the Gulf of Aden. ZOHARY (1973) described this region as "Eritreo-Arabian Province" within the "Sudanian Floristic Region" according to its high floristic conformity; NEWTON (1980) used the term "Mandab Circle".

In mainland Africa, the taxa of sect. *Arabica* are restricted to the "Somali-Masai-Region" sensu WHITE (1983), which encloses the "Eritrean Region" sensu ZOHARY (1973). The "Somali-Masai Region", which houses a high percentage of endemics, reaches its northernmost distribution in the Etbai Mountains of southern Sudan (ca. 19° N). This is also the locality where sect. *Arabica* with *D. sulcata* on the Erkowit plateau has its northwestern distribution limit. Except for *D. sulcata* subspp. *sulcata* and *seminuda,* all taxa of this region inhabit only small areas. The countries surrounding the southern Red Sea lodge a single taxon or species each, except Somalia with three species, two of them being endemics that have been collected just once or twice.

Duvalia sulcata represents a characteristic element within the succulent-rich, southwesterly exposed sites of the Tihama, the Escarpment and the Asir (Saudi Arabia, Yemen) and the southern Mountains of the Yemen including the Hadramaut, which is the floristically richest part of Arabia (cf. AL-HUBAISHI & MÜLLER-HOHENSTEIN 1984).

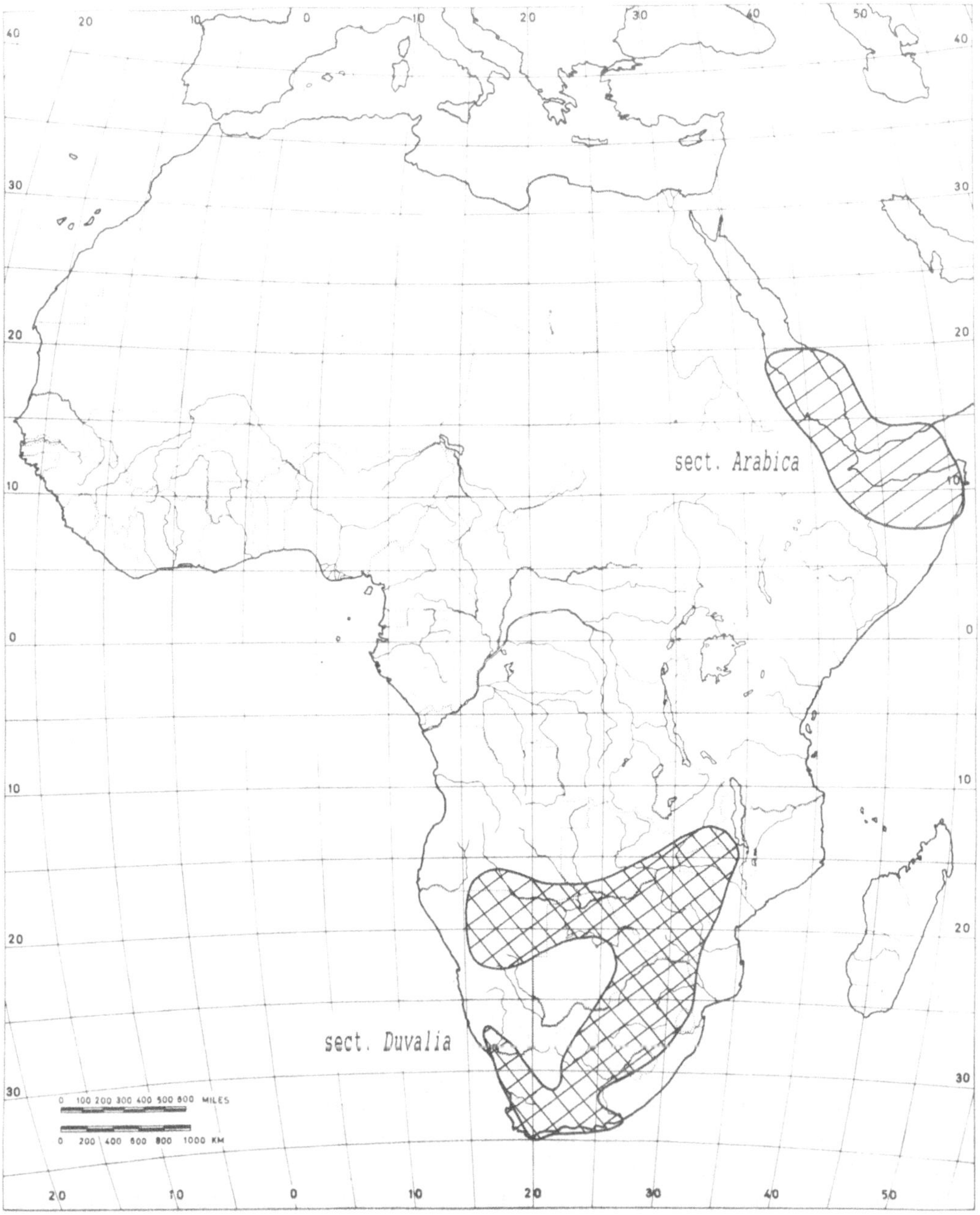

Fig. 1. Distribution of *Duvalia* sectt. *Duvalia* (cross-hatched) and *Arabica* (hatched)

The close floristic relationships between SW Arabia and Africa is demonstrated by the high number of Eritreo-Arabian geoelements. Many species of herbs, woody or succulent plants display disjunct distribution patterns similar to that encountered in *D. sulcata* (see DEIL 1988a for examples). However, there exist also connections of this region to the "Nubic-Sindian Province" (cf. KÜRSCHNER 1986).

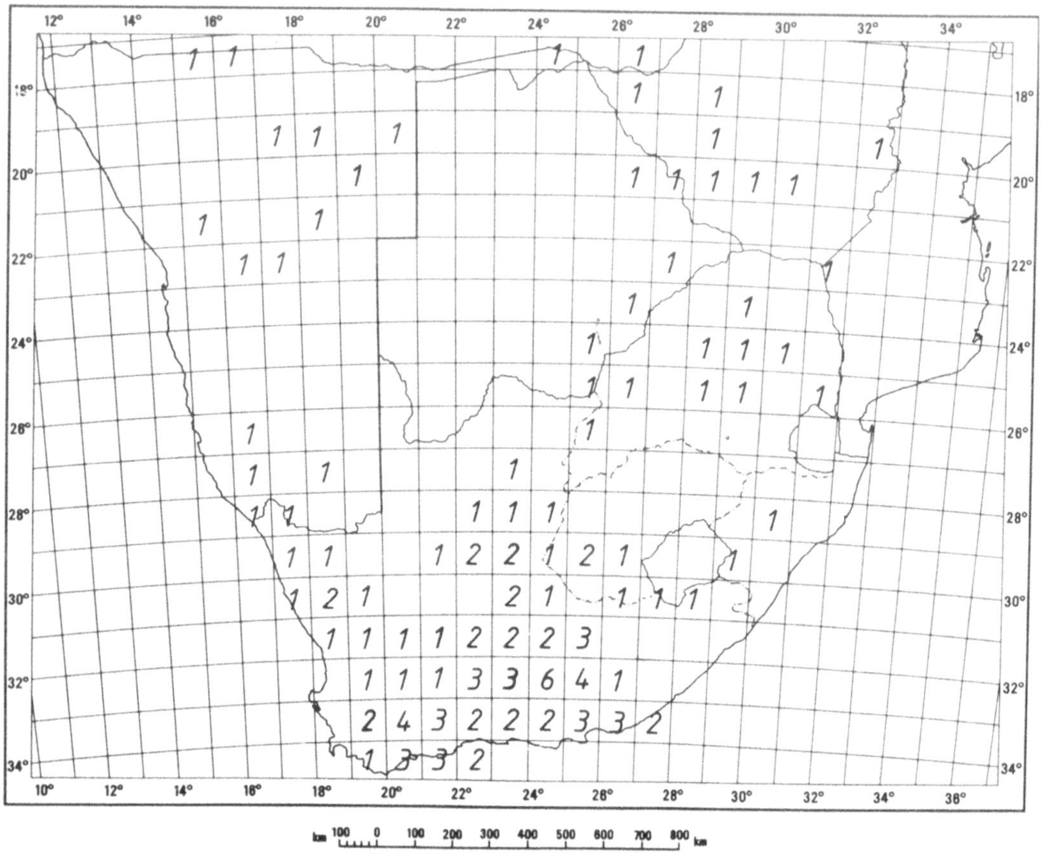

Fig. 2. Distribution and species concentration of *Duvalia* sect. *Duvalia* in southern Africa

Habitats. Climate. Apart from fog and dew, the average precipitation at *Duvalia* localities in Namibia and South Africa amounts to ca. 100–300 mm/p.a., and the average summer temperatures rarely reach more than 25–30 °C. The maximum temperatures, however, may reach 45 °C. During winter not all *Duvalia* localities are free of frost. At some higher sites of the Western Mountain Region up to minus 10 °C frost can be expected.

Duvalia polita is the only species that is restricted to the frost free summer rainfall areas of southern Africa.

The climate in Arabia is strongly influenced by summer or winter rainfall, sometimes superposed by altitudinal rain in the escarpment and the Asir respectively (König 1987). Relief and altitude are the dominant parameters for the distribution of precipitation, apparently more important than in southern Africa. The semi-arid *Duvalia* localities receive 100–400 mm rain p.a. (occasionally more at higher sites of the western escarpments and at the mountains of the southern Yemen), but precepitation often is erratic and local. The average temperature during the year is around 30 °C, while the maximum temperatures can exceed 45 °C (cf. König 1987, Deil 1988b).

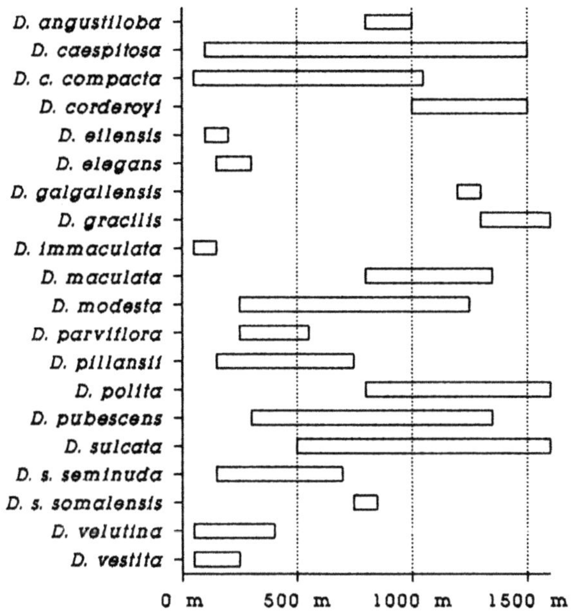

Fig. 3. Altitudinal ranges of *Duvalia* habitats

Nearly temperate climatic conditions dominate in the disjunct Sudanian area of *D. sulcata*. Here, it is comparatively cool due to the relatively high altitudes (ca. 1200 m a.s.l.) and the many foggy days. The annual precipitation is 180–220 mm on average, mainly falling as a combination of summer and winter rainfall (cf. KASSAS 1956).

Exposure and soil. Relief and exposure influence the habitat preferences of the species. Typical habitats are westernly exposed lower slopes of hills and mountains with 1 to 5 degree inclination. Such localities in Namaqualand and the southern Namib also receive more fog. The soil is predominantly rich in particles (loam to sand) and slightly acid at pH 5 to 7(8).

Microhabitats of South African species. Like many stapeliads and other succulents in southern Africa (e.g., small-sized *Senecio, Pelargonium, Euphorbia, Gasteria, Haworthia, Aizoaceae* and many more), *Duvalia* species have developed a successful settling and survival strategy. About 90% of the plants observed in their habitat were growing under shrubs. The "host plants" of stapeliads ("Vasallenpflanzen", MARLOTH 1908) are usually shrubs while stem or leaf succulents can rarely be found between them. This narrow niche in a semi-arid environment offers considerable benefit for the hidden partner, because the host plants accumulate detritus at their base and offer protection against grazing, erosion and direct radiation. Thus they provide a favourable microclimate.

Duvalia sect. *Duvalia* as typical element of Karoo and Savannah vegetation types. Considerable altitudinal preferences are obvious (Fig. 3). Widely distributed species, e.g., *D. caespitosa* and *D. pubescens*, are characterized by correspondingly wide ecological amplitudes. Other species are more dependent on specific ecological conditions (*D. angustiloba, D. elegans*, see species descriptions), or are characteristic of karroid highlands (*D. corderoyi, D. gracilis, D. maculata* p.p.). In the coastal lowlands of the southern Western

Table 1. Phytosociological observations on a *Duvalia pubescens* locality with 11 individuals. (Species list arranged according to life forms and abundance/dominance. Abundance/dominance assessments follow the method of BRAUN-BLANQUET 1964: *r* 1 individual/investigation area; + 2–5 individuals/investigation area; *1* 6–50 individuals/investigation area; *2a* any individual number/investigation area, coverage 5–15%; *2b* coverage 16–25%; *3* coverage 26–50%. Succulents are succulents except shrubby/leafy forms; *AVT* Acocks Veld Type)

Date:	31 Oct. 1986
Locality:	8.5 km N Concordia (Namaqualand, Grid:2917 DB)
AVT:	33, Namaqualand Broken Veld
Exposure:	northwest
Inclination:	ca. 2 %
Altitude:	1100 m
Soil:	quartzic sand with quartz pebbles on granite
Area:	25 × 50 m
Coverage, total:	15 %
" shrubs:	12 %
" succulents:	2 %
Height :	1 – 50 cm

Remarks: Vegetation coverage dominated by shrubs, though coverage is sparse with many shrubs severely grazed, erosion starting

Species:

Ruschia robusta L. BOLUS	2a
Tetragonia fruticosa L.	1
Odonthophorus nanus L. BOLUS	1
Ruschia sp.	1
Drosanthemum hispidum (L.) SCHWANTES	1
Galenia cristallina (ECKL. & ZEYH.) FENZL	1
Galenia africana L.	+
Hermannia disermifolia JACQ.	+
Euphorbia cf. *rhombifolia*	+
Lampranthus hoerleinianus (DINTER) FRIEDR.	+
Euphorbia decussata E. MEYER ex BOISS.	+
Othonna cf. *obtusifolia*	+
Pentzia suffruticosa (L.) HUTCH. ex MERXM.	+
Euphorbia tuberculata JACQ.	r
Sphalmanthus scintillans (DINTER) DINTER & SCHWANTES	+
Mesembryanthemum pellitum FRIEDR.	+
Crassula namaquensis SCHONL. & BAK.f.	+
Crassula muscosa L. subsp. *muscosa*	+
Pelargonium carnosum (L.) L'HÉRIT.	+
Pelargonium sp.	+
Quaqua acutiloba (N. E. BR.) BRUYNS	+
Senecio niveus (THUNB.) WILLD.	+
Leysera gnaphalodes (L.) L.	1
Ifloga polycnemoides FENZL.	1
Pharnaceum ciliare ADAMSON	1
Hirpicium alienatum (THUNB.) DRUCE	1
Hypertelis salsoloides (BURCH.) ADAMSON	1

Table 1 (continued)

Spergularia media (L.) PRESL.	1
Gorteria diffusa THUNB.	+
Arctotis sp.	+
Amphiglossa tomentosa (THUNB.) HARVEY	+
Bromus molliformis LLOYD	+
Helichrysum cf. *obtusum*	+
Ursinia sp.	+

Table 2. Phytosociological observations on a *Duvalia pubescens* locality with five individuals. (For further explanation see Table 1)

Date:	2 Nov. 1986	
Locality:	3 km E of Kamieskroon (S Namaqualand,	
AVT:	33, Namaqualand Broken Veld (Grid 3017 BB)	
Exposure:	–	
Inclination:	–	
Altitude:	800 m	
Soil:	sand	
Area:	10 × 25 m	
Coverage, total:	40 %	
Coverage, shrubs:	35 %	
Coverage, succulents:	5 %	
Height:	1 – 70 cm	
Remarks:	no Mesembryanthema	
Species:		
Galenia africana L.		3
Euphorbia mauretanica L.		2a
Eriocephalus ericoides (L. f.) DRUCE		2a
Pentzia incana (THUNB.) KUNTZE		1
Pentzia suffruticosa (L.) HUTCH. ex MERXM.		1
Thesium hispidulum LAM. ex SONDER var. *subglabrum* A.W. HILL		+
Didelta spinosa (L. f.) AITON		+
Relhania genistifolia (L.) L'HÉRIT.		+
Pelargonium praemorsum (ANDR.) F. DIETR.		r
Pelargonium angustipetalum sp. nova MARAIS ined.		r
Aloe variegata L.		+
Tylecodon wallichii (HARVEY) TOELKEN		+
Quaqua incarnata (L. f.) BRUYNS s. str.		+
Stapelia acuminata MASSON		+
Gorteria diffusa THUNB.		1
Hirpicium alienatum (THUNB.) DRUCE		1
Osteospermum hyoseroides (DC.) NORLINDH		+
Ehrharta barbinodis NEES ex TRIN.		r
Tribolium hispidum (THUNBG.) RENVOIZE		r

Table 3. Phytosociological observations on a locality of both *Duvalia elegans* (three individuals) and *Duvalia caespitosa* var. *caespitosa* (one individual). (For further explanation see Table 1)

Date:	23 Nov. 1986
Locality:	Langvlei, 10 km NW Robertson (Grid:3319 DD)
AVT:	26, Karroid Broken Veld
Exposure:	southwesterly exposed base of hill
Inclination:	2–3 %
Altitude:	300 m
Soil:	firmed, coarse sand and above 20 % boulders
Area:	10 × 10 m
Coverage, total:	90 %
Coverage, shrubs:	90 %
Coverage, succulents:	5 %
Mosses & lichens:	ca. 50 % coverage
Height:	5 – 120 cm

Remarks: extremely dense, partly matted vegetation, rich in small succulents; the undisturbed situation observed and the high coverage with lichens and mosses is obviously due to the absence of grazing during the last year(s)

Species:	
Pentzia incana (THUNB.) KUNTZE	3
Pteronia incana (BURM.) DC.	3
Crassula perfossa (HARVEY) MARLOTH	2b
Euphorbia burmannii E. MEYER ex BOISS.	2a
Euphorbia mauretanica L. var. *mauretanica*	1
Crassula subaphylla (ECKL. & ZEYH.) HARVEY	1
Euphorbia nesemannia R. A. DYER	+
Drosanthemum speciosum (HAW.) SCHWANTES	+
Drosanthemum sp.	r
Tetragonia cf. *robusta*	+
Eriocephalus ericoides (L. f.) DRUCE	+
Galenia africana L.	+
Galenia secunda (L. f.) SONDER	+
Salsola aff. *aphylla*	+
Ruschia caroli (L. BOLUS) SCHWANTES	+
Aridaria noctiflora (L.) SCHWANTES	r
Euryops tenuissimus (L.) DC. subsp. *tenuissimus*	+
Aizoon sp.	r
Crassula muscosa L.	1
Delosperma cf. *dejagerae*	+
Cotyledon orbiculata L.	+
Senecio radicans (L. f.) SCH. BIP.	+
Mesembryanthemum sp.	r
Tylecodon paniculatus (L. f.) TOELKEN	+
Adromischus filicaulis (ECKL. & ZEYH.) C. A. SM.	+
Anacampseros telephiastrum DC.	+
Hirpicium integrifolium (THUNB.) LESS.	+
Ehrharta calycina J. E. SM.	+
Karroochloa curva (NEES) CONERT & TURPE	r

Cape, *D. immaculata* and *D. vestita* prefer Fynbos-like vegetation, particularly Coastal Fynbos, which is usually dominated by small succulents and grasses, or Renosterveld.

In Namibia and Zimbabwe, *D. polita* is found preferably in *Acacia*- and Mopane bushland. Most of these localities are subject to regular fires during the dry season. The presence of subterranean stems (stolons, see morphological section) can be interpreted as adequate adaptations.

Most taxa of sect. *Duvalia* are indicator species of karroid vegetation of around 15–40% plant coverage.

One quarter of all specimens investigated was collected in a single veld type (Acocks 1988), the "Karroid Broken Veld". In contrast, in grassy highlands (± 1500 m) just a few "runaways" can be found. The Little Karoo (Robertson, Montagu, Worcester, Ladismith) owes its species richness to the small-scale intercalation of semi-arid and mediterranean elements caused by highly variable relief-dependent precipitation. Leaf and stem succulents of variable systematic affinities dominate here, receiving 150–300 mm annual precipitation. For demonstration a study of the vegetation of three *Duvalia* localities from Namaqualand and the Little Karoo is presented (Tables 1–3). However, the species lists resulted from a single visit and a fair number of geophytes may have been missed.

Duvalia sect. *Arabica* as an element of semi-arid African/Arabian vegetation types. *Duvalia* habitats are found in particular in two of the nine vegetation units described by White (1983) for the "Somali-Masai regional centre of endemism": (a) Somali-Masai *Acacia-Commiphora* dry scrub, and (b) semi-desert grass and scrubland.

A more detailed description of the locality for plagiotropic *Stapelieae* in southern Sudan is given by Kassas (1956). Near Erkovit, a locality of *D. sulcata*, an average cover of 20–30% was observed. It was dominated by *Euphorbia abyssinica*, *Acacia* spp. and various shrubs. According to Collenette (1985, and pers. comm.) and König (1987) the Saudi-Arabian habitats do not differ from the African ones, in which stapeliads are often encountered in *Acacia-Commiphora* dry scrub.

Succulent plant communities in N Yemen have been studied by Deil (1988b). Communities rich in stapeliads occur in southwestern Arabia mainly on the west-exposed slopes of the Tihama and the escarpment, between 250 and 2000 ma.s.l. At higher altitudes, as high as the peaks, succulent *Asclepiadaceae* and *Euphorbiaceae* are replaced by *Crassulaceae* and *Aizoaceae* (*Delosperma*; Deil & Müller-Hohenstein 1988). Succulents frequently occur as elements of anthropozoogenously influenced vegetation, a situation described by Deil (1988b) for the ruderal Caralluma subulatea – Euphorbietum inarticulatae – association. *Duvalia sulcata* represents a character species of this replacement vegetation below 500 m a.s.l.

Morphology

Stems. All species have succulent, plagiotropous stems that form prostrate, compact mats. Sometimes, these consist of more than a hundred single stems covering half a square metre or more. In a few species additional subterranean stems (stolons) are produced (Fig. 41C).

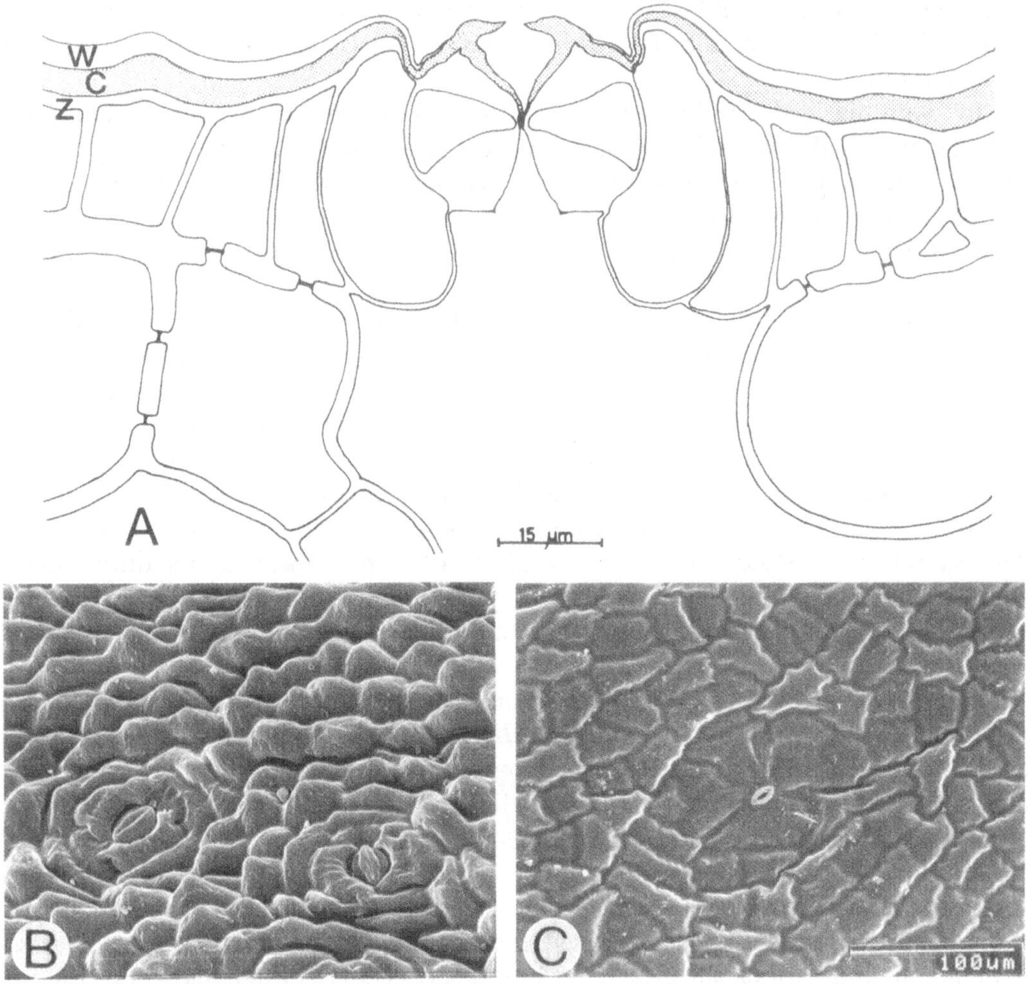

Fig. 4. Stem epidermis of *Duvalia* species. *A Duvalia polita*, cross section (*w* wax layer, *c* cuticle, *z* cellulose wall; LM); *B D. caespitosa* var. *caespitosa*, K 1146, epidermal surface (SEM); *C D. sulcata* subsp. *seminuda*, Barad 11723, epidermal surface (SEM)

The green to grey-green stems are 4- to 6-angled and ovoid, cylindrical to clavate in shape. The single stem measures one to several cm in length and 1 to 2 cm in diameter. The stolons (rhizomes) usually surpass the green stems in length, force themselves through the soil and penetrate its surface a short distance from the mother plant. These stolons are ivory-coloured, rarely thicker than a pencil, and beset with scanty leaf rudiments.

All species bear small rudiments of leaves on raised tubercles (podaria). They are arranged in decussate order, only in the strictly 6-ribbed species (*D. corderoyi, D. polita*) they appear along the six orthostichies. The length of the leaf rudiments is often a specific character, for instance in *D. parviflora* they do not exceed 1 mm in length, while in *D. maculata* and *D. polita* they may measure up to 6 mm. In sect. *Duvalia* the triangular to lanceolate leaves show a clearly dorsiventral structure. Being green in a primary stage, the leaves wither soon. They may remain on the podaria as dry scales, or, more frequently, fall off. A corky abscission tissue between the stem and the leaf is responsible for the leaf shedding (ALBERS & al. 1989). In sect. *Arabica*, in contrast, the leaf rudiments are terete, subulate and more or less persistent.

The leaf rudiments (as well as the floral bracts) are at their base accompanied by small and weak globular structures, whose nature as stipular glands has been demonstrated by MEVE & ALBERS (1990a). However, they are strongly reduced or even lacking in *D. immaculata* and *D. vestita*.

In sect. *Arabica* stipules are totally absent on the vegetative body, but can be found occasionally on the bracts.

Stem epidermis. The stem epidermis of stapeliads has been studied in detail by ZEMKE (1939), STRAKA 1979, ALBERS & al. (1989), MEVE & al. (1990). The epidermis is monolayered and consists of isodiametric cells (Fig. 4). The outer cell wall shows a structure characteristic of succulents with a thick cellulose wall, a thin, pectin-rich subcuticular layer, which is only traceable in the electron microscope, a reticulate cuticle of ca. 5 µm in diameter and a smooth wax layer measuring 3–4 µm in thickness (Fig. 4A, see also MEVE & al. 1990). Sporadically wax crystalloids are found on top. The stomata are only slightly sunken as compared to other plants of arid habitats.

The general description of stapeliad stem surface structures (see ALBERS & al. 1989) is to be supplemented for *Duvalia* as follows: (1) the basal shape of epidermal cells is isodiametrical-polygonal, (2) the edges of epidermal cells are inconspicuous, (3) the surface of wax lamella is more or less glabrous.

The micromorphology of the periclinals allows the division into two groups that correspond to the taxonomic subdivision of the genus:

1. Periclinals slightly convex to (rarely) cone-shaped, often with central papilla, anticlinals straight (Fig. 4B): sect. *Duvalia*. The most pronounced sculpturing is found in *D. caespitosa* and *D. pubescens*.

2. Periclinals very slightly convex, without central papilla; anticlinals occasionally slightly undulated (Fig. 4C): sect. *Arabica*.

Inflorescences. Inflorescence bracts. The inflorescence formation starts with the development of a single bract, which is associated with the terminal flower. Two bracteoles can be found with each of the successive flowers. The tiny, chlorophyll-free leaflets normally bear stipular glands (MEVE & ALBERS 1990a).

Inflorescence structure. In contrast to a number of orthotropous *Stapelieae* (e.g., *Rhytidocaulon, Pseudolithos*), *Duvalia* never develops inflorescences on the primary shoot. The plants flower at the earliest in the second year after germination and after the termination of primary growth of the innovation shoots of higher order. In the *Stapelieae* the usually extra-axillary positioned synflorescences are always terminated by a flower. They are therefore determinate (closed) as is typical for all *Asclepiadaceae* (cf. LIEDE & WEBERLING 1995). The shoot axis is a sympodial structure and can be attributed to the monotelic synflorescence type (TROLL 1964). Here as well as in most of the advanced *Asclepiadaceae*, the synflorescences are placed in a lateral position by the vegetative stem, which continues growth from one of the axilliary buds (Fig. 5), resulting the displacement of the synflorescence into the extra-axillary position (DEMETER 1922, TROLL 1959).

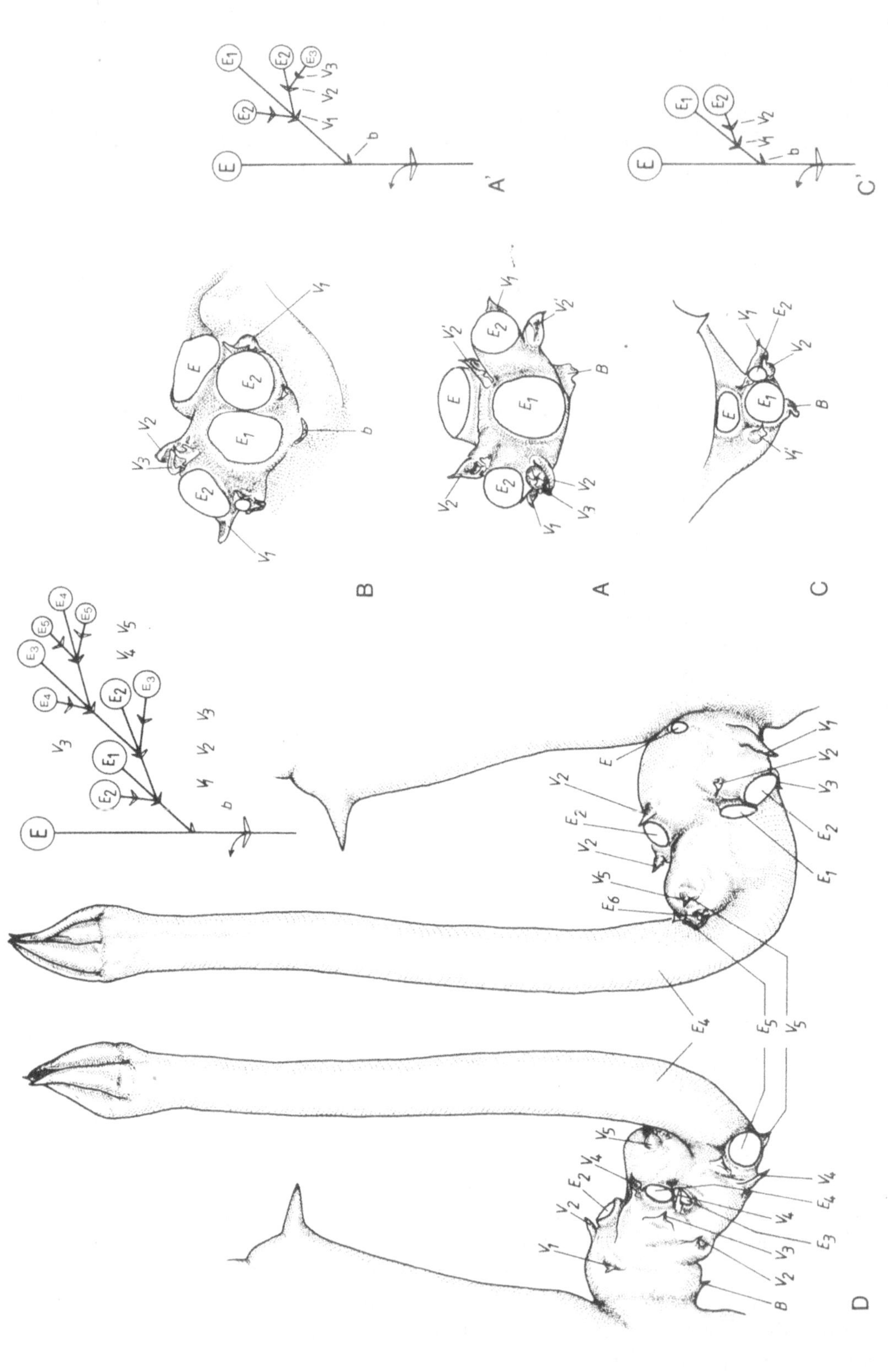

Fig. 5. *Duvalia* synflorescences. *A, A' D. elegans* (Albers & al. *sub* K 1383); *B D. caespitosa* var. *caespitosa* (Meve 252); *C, C' D. immaculata* (Meve 399); *D, D. sulcata* subsp. *sulcata* (Hartmann 21506). (B bract, V bracteole, E flower)

Duvalia is characterized by more or less reduced cymoid-thyrsoid inflorescences, which are derived from dichasia or monochasia (Fig. 5; compare also WERTEL 1976, LIEDE & WEBERLING 1995). Young inflorescences frequently develop a dichasium at the first branching of the sympodium (Fig. 5A, B, D). In older inflorescences only one of the two bracteal buds is stimulated (exceptions are not rare, see below). The other, unstimulated bud is used to form just a single terminal flower; the resulting inflorescence structure is a bostrychoid (screw-like) cyme (Fig. 5A, A′). This inflorescence type is found in many other representatives of *Stapelieae* (e.g., *Ceropegia*; TROLL 1959, BRUYNS 1985, LIEDE & WEBERLING 1995).

As in *Huernia* (WERTEL 1976), there are inflorescences that show a distinct and continuous dichasial arrangement of their equally dichasial partial inflorescences (e.g., in *D. angustiloba*). This results in the formation of a succulent rachis, carrying the actually flowering inflorescences like a stalk (Fig. 5D). In *D. angustiloba* the rachis is conspicuously dorsiventrally flattened. In *D. sulcata*, in contrast, it is more rounded due to the bostrychoid partial inflorescences (Fig. 5D). The rachis can be long-lived while flowering activity continues for several years and elongates with every newly developed inflorescence. Finally, extremely reduced inflorescences, in which only single terminal flowers develop, can be found (Fig. 5C, C′).

Flowering times and cycles. Flower formation shows a clear periodicity (Fig. 6). In almost all species, main flowering seasons can be observed, during which almost all individuals flowers. Deviations, however, are frequent, and 'too' early or late flowers occur. Flowering under cultivation in the greenhouse in Münster starts usually in May with the relatively small-flowered species *D. angustiloba, D. caespitosa* var. *compacta, D. gracilis*, *D. modesta* and *D. parviflora*, and extends into late November. Only *D. parviflora* has been observed to flower almost all year round (Fig. 6). Field observations and literature data (BOND & GOLDBLATT

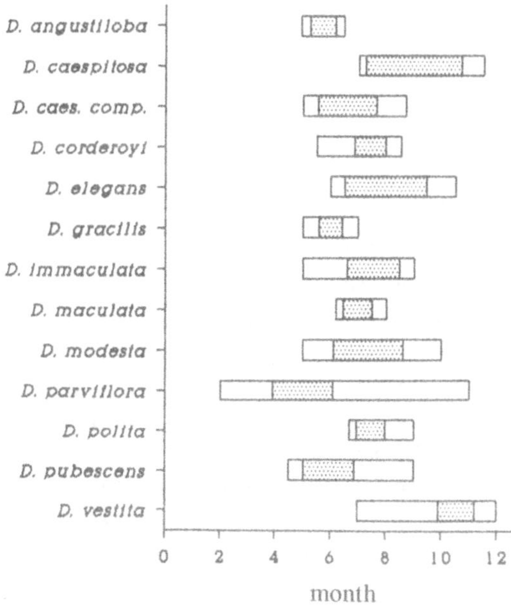

Fig. 6. Flowering times of species of *Duvalia* sect. *Duvalia* in the greenhouse at Münster Botanic Garden, Germany

1984) show that the natural flowering times in South Africa are from early summer to autumn. This means that there is a six month shift in the flowering season after removal of the plants from their natural habitats, placing it again to (European) summertime. Thus, the induction of flowering is obviously caused by the light factor. Long days, or days with declining length seem to induce flowering, and the inductive day length seems to be species specific and genetically controlled. This impression is reinforced by the distribution pattern shown in Fig. 2, which shows some correlation between flowering season and distribution. Corresponding to seasonality of rainfall in the Western Cape Province, the main flowering seasons of some species of the Northern and Eastern Cape Province are retarded under the uniform cultivation conditions in the greenhouse.

Light-induced flowering is also known from other succulents in southern Africa, such as the *Aizoaceae-Ruschioideae* (*Conophytum, Lithops*: S. HAMMER, pers. comm.; *Pleiospilos*: S. LIEDE, pers. comm.), or *Crassulaceae*, for instance *Tylecodon wallichii* (HARV.) TÖLKEN.

The duration of flowering is highly variable. The long flowering season of the species with the smallest flowers, *D. parviflora*, might be interpreted as a compensation for the probably low attractivity of its flowers to pollinators. In addition, there is a clear tendency for those species with modest flowers (e.g., *D. angustiloba*, *D. gracilis*) to compete successfully for pollinators against stapeliads possessing more showy flowers by starting to flower early in the season.

Flower morphology. Like all stapeliad flowers, *Duvalia* flowers are radially symmetrical and are clearly divided into a calyx, a corolla and a gynostegium with a double corona.

Pedicel. Flowers are normally carried on a 10–25 mm long, glabrous, horizontal pedicel that bends into a vertical position just below the calyx, thus presenting the flowers in a horizontal position. The flowers are normally formed on young shoots, and thus situated at the periphery of the plant. Also long pedicels (up to 7 cm long in *D. velutina*) optimizes flower presentation by removing the flower from the tangle of branches and leaves of the "host plant" onto a more strongly contrasting and less shady background.

Calyx. The calyx consists of five free sepals and is morphologically relatively uniform. The sepals are green, semi-succulent, triangular, acuminate, and measure a few millimeters. Species of sect. *Arabica* may have up to 6 mm long, lanceolate sepals.

Adaxially, in the sinuses between the sepals, glandular emergences (colleters) are found as in all stapeliads (MEVE & ALBERS 1990a).

Corolla. The corolla is radially symmetrical, five-merous, and fused for at least one fifth to at the most one half of the flower diameter. Main parts of the corolla tube are folded to form a fleshy ring (annulus). This annulus is of the corolline corona type (cf. LIEDE & KUNZE 1993). It harbours the stalked gynostegium including the gynostegial corona and can be almost completely covered by the coronal disc (e.g., *D. elegans, D. sulcata* s.str.; Fig. 7A, C). The coronal disc normally closes the annulus along its interior margin. Basally, at the transition to the free corolla lobes, the annulus is often constricted (Fig. 7B). A few species (*D. elegans, D. sulcata* subsp. *sulcata*) possess flat annuli; in others (*D. modesta, D. pillansii, D. polita, D. eilensis, D. velutina*) the annuli lack basal

Fig. 7. *Duvalia* flowers in longitudinal sections. *A D. elegans*; *B D. maculata*; *C D. sulcata*. [*A* anther, *An* annulus, *AW* anther wing, *Ca* carpel, *Cs* staminal corona lobe, *C(is)* coronal disc, *NC* nectar cavity, *P* pollinium, *SH* stylar head]

constrictions and emerge gradually into the free corolla lobes (Fig. 7C). Annuli occur frequently in *Huernia, Orbea* and *Stapelianthus*, the three genera most closely related to *Duvalia*, and less prominently in some other stapeliad genera.

The free corolla lobes are normally much longer than broad (exceptions occur in *D. velutina*), narrowly to broadly triangular and acute. The blades are flat and spreading or bent outward to a variable degree. Mostly, however, the two halves of the blade are reflexed so strongly along the midrib that their margins almost touch along the abaxial corolla surface ("*D. reclinata*"). In both sections, weak to strong longitudinal furrows may sculpt the corolla lobes (*D. gracilis*, Fig. 8A; *D. sulcata* subsp. *seminuda*, Fig. 8B).

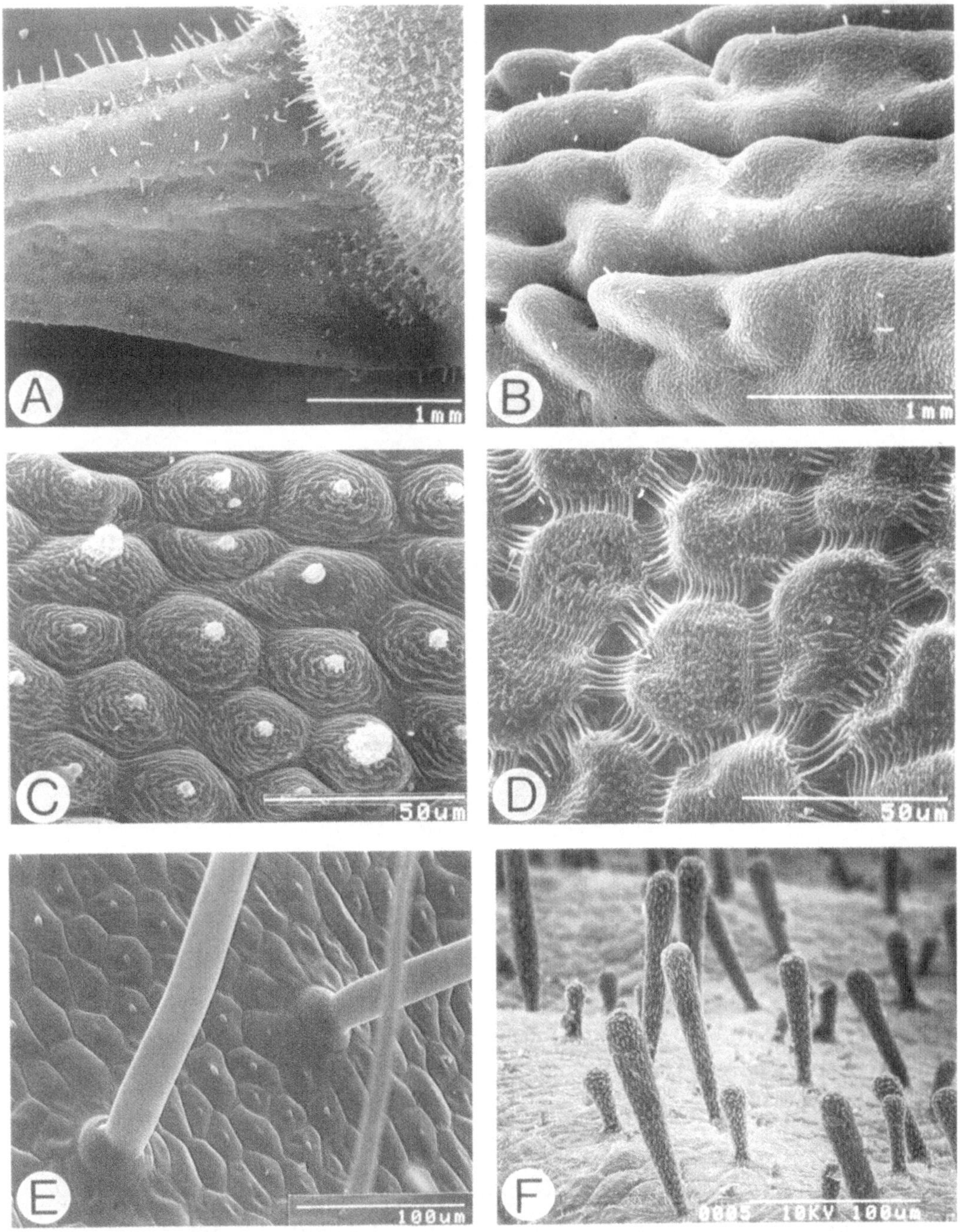

Fig. 8. Corolla in *Duvalia* species, SEM. *A* rim of annulus with basal half of corolla lobe (*D. gracilis,* MEVE 346); *B* corolla lobe surface, medium section (*D. sulcata* subsp. *seminuda*, BARAD 11729); *C* epidermal surface of basal corolla lobe (*D. caespitosa* var. *caespitosa*, MEVE & KUSCH sub K 1550); *D* epidermal surface of corolla lobe (*D. sulcata* subsp. *seminuda*, NOLTEE 1898); *E* epidermal surface with corolla lobe hair bases (*D. elegans*, MEVE 394); *F* obconical epidermal hairs of corolla lobe (*D. velutina*, DEIL 1240)

Corolla epidermis. The basal epidermal pattern of the adaxial corolla and its indumentum can be described as follows: (1) basal cell pattern homogeneous, (2) shape of cells isodiametrical-polygonal, (3) anticlinals ± straight, (4) anticlinal boundaries slightly sunken, (5) periclinals (tabular-)convex, partly with central papilla, and/or central sector with cuticle usually finely curled, (6) anticlinal sector with cuticle parallelly pleated.

In addition, papillae (=ca. 20–150 μm long), hair papillae (=ca. 150–250 μm long), and hairs (more than 250 μm long) can be found on the adaxial corolla surface, as well as vibratile clavate hairs along the margins.

The two sections can be differentiated by their epidermal structure and sculpture. In sect. *Duvalia*, convex periclinals with central papillae (Fig. 8C) are commonly found. The cuticles show homogeneously sculptured surfaces. Papillae and hairs are typically conical. In contrast, in sect. *Arabica* the corolla epidermis has smaller cells with flat or only slightly convex periclinals, mostly lacking a central papilla (Fig. 8D). The cuticular surfaces are usually divided into a finely curled central sector and a roughly pleated anticlinal one. The shape of hairs is diverse.

At species level, many taxa can be recognized by their epidermal features. Macroscopically brilliant or dull flower surfaces reflect microscopic characters, e.g., smooth cuticles are responsible for brilliant flower surfaces of *D. elegans* and *D. polita*, and an intensely pleated cuticle for the dull flower surface of *D. immaculata*. Hair shape is species-specific in some species of sect. *Arabica* (e.g., the obconical hairs of *D. velutina*, Fig. 8F).

Clavate hairs along the corolla margins occur in both sections. They vary in length, but scarcely in shape and sculpture. All kinds of corolla hairs observed have a finely curled to verrucose cuticular surface (cf. Fig. 8E, F).

The pollination apparatus. The pollination apparatus in the *Asclepiadaceae* s. str. consists of the gynostegium, the central product of fusion of the stylar head with the filaments and the anthers. In all *Stapelieae*, the gynostegium is supplemented by coronal lobules to complete the compact reproduction unit.

In *Duvalia*, the gynostegium is stipitate (placed on top of a column) – a strongly elongated 'basal tube', mainly of receptacular origin, lifts the filament tube proper (Fig. 7). The five guide rails, each formed by two adjacent anther wings, are vertically to obliquely oriented at the flanks of the filament tube (Figs. 7, 9A). The sclerified and therefore cartilaginous margins of the guide rails spread in a wing-like manner in the area of their broadened bases (Fig. 9A, B). Narrowing continuously, they lead up to the translator. The inner surfaces of the anther wings are lined with upwardly oriented glandular bristles (Fig. 9B), which obviously serve as a system promoting the introduction of insect hairs or pellucid pollinium margins into the guide rails, or rather prevent their downward withdrawal. The guide rails have a simple structure and are not differentiated (in cross-section) into an outer groove and an inner tube (as, for instance, in *Huernia*, cf. Kunze 1982). Below the guide rails, sunken into the filament tube, spacious nectar cavities occur with small, circular to slit-like openings (Fig. 9A). Glandular epithelia line these nectar cavities, but secrete only little nectar.

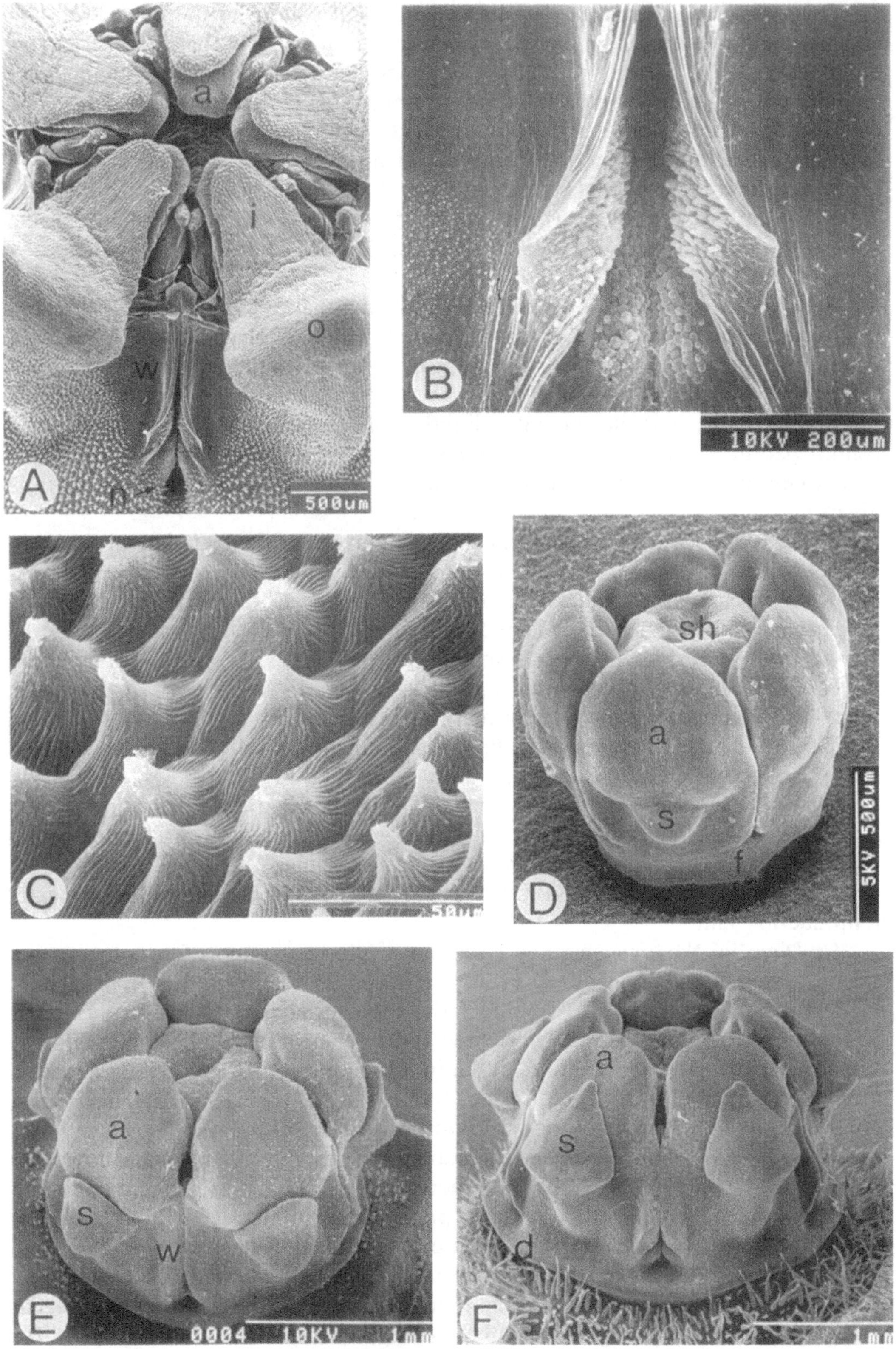
a
i
o
w
n
A
500um
B
10KV 200um
C
50um
sh
a
s
f
D
5KV 500um
a
s
w
E
0004 10KV 1mm
a
s
d
F
1mm

The stylar head does not show any peculiarities in comparison to those of other stapeliad genera (Fig. 7). The white, discoid structure is pentagonal, adaxially convex, usually with a slight central depression. Abaxially, the postgenitally fused tips of the carpels intermediate the contact to the concave central region of the stylar head.

Corona. The corona of the *Stapelieae* consists of outgrowths of the backs of the stamens (staminal corona) and the filament tube (interstaminal corona; SCHUMANN 1895, KUNZE 1982, 1990). In *Duvalia* the corona is double, separated into a staminal corona (Cs) and a ring-like structure (coronal disc) formed by the fused interstaminal corona and the basal parts of the staminal corona [C(is), Fig. 7; for corona terminology and abbreviations see LIEDE & KUNZE 1993]. The staminal corona is apically differentiated into a horizontal inner and an erect outer appendage. The shape of the coronal disc is not always circular, but rather pentagonal or even decagonal, and in *D. angustiloba* and *D. parviflora* it constitutes a thickenend fringe rather than a disc.

The epidermal cells of the staminal and interstaminal coronas possess different patterns and surfaces. The Cs lobules are characterized by flat epidermal cells of isodiametrical outline, often covered by a considerably pleated cuticle. The epidermis of the coronal disc, in contrast, shows a higher degree of sculpturing with elongated central papillae crowning mostly elongated cells (Fig. 9C). The cells are arranged in "rows", which lead from the outer rim of the coronal disc to the gynostegium. Often, epidermal cells of the corona are active as osmophores.

Ontogeny of the corona. Scanning electron microscope studies (Fig. 9D–F) confirm previous ontogenetic studies (KUNZE 1982, 1990, HOFMANN & SPECHT 1986) according to which the formation of the corona is a secondary process. The gynostegium with the large anthers can be recognized long before the first coronal structures appear in the buds. The gynostegium is by no means fully differentiated, but the anther wings and parts of the filament tube with the preformation of the guide rails has started and the stylar head is mightily developed (Fig. 9D). It is only at this stage that the formation of the corona starts with the swelling of a small meristem hump at the base of the back of the anthers. Shortly afterwards, but clearly delayed, the meristem ring of the coronal disc starts to develop (Fig. 9E, F).

The thickening of the corona parallels the intercalary elongation of the filament tube between the guide rail entrance and the coronal disc, so that a separation of the two coronas takes place. In this late stage of floral development the basal sterile part of the filament tube, or the receptacle, elongates below the corona base parallel to the folding of the annular corolline corona (annulus, Ca).

Glandular fields are situated at the top of the interstaminally protruding tissue stripe of the stylar head. Shortly before the opening of the flower the palisade cells

---

Fig. 9. Corona of *Duvalia* species, SEM. *A* central part of the pollination apparatus (*D. caespitosa*); *B* basal half of guide rail lined with glandular hairs (*D. sulcata* subsp. *sulcata*, HARTMANN 21506); *C* epidermal surface of coronal disc (*D. polita*, ALBERS & al. 544); *D–F* ontogeny of the pollination apparatus (*D. pubescens*). (*a* anther, *d* coronal disc, *i* inner part of staminal corona lobe, *o* outer part of staminal corona lobe, *n* entrance to nectar cavity, *f* filament tube, *s* staminal corona, *sh* stylar head, *w* anther wing)

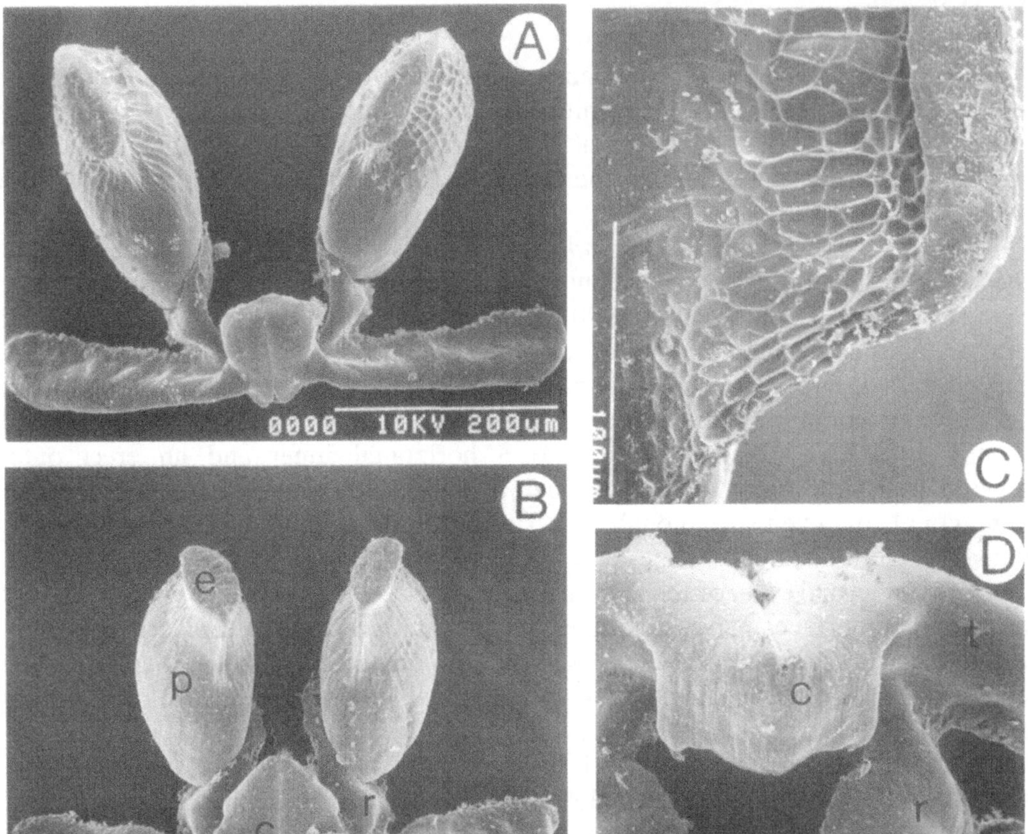

Fig. 10. Pollinarium in *Duvalia* species, SEM. *A* Pollinarium (*D. angustiloba*, Meve 321); *B* pollinarium (*D. sulcata* subsp. *seminuda*, Noltee 1898); *C* apical part of pellucid margin (*D. sulcata* subsp. *seminuda*, Noltee 1898); *D* fastening of the caudicle to the corpusculum in top view (*D. sulcata* subsp. *sulcata*, Hartmann 21507); *E* fastening pollinium base/caudicle (*D. pubescens*, Meve 230); *F* pollinarium of *D. angustiloba* attached to the labellum of proboscis of *Musca domestica*. (*c* corpusculum, *e* pellucid margin, *p* pollinium, *r* caudicle, *t* translator wing)

of that tissue exude the corpuscular secretions into a preformed cavity between the anther margins and the stylar head (DEMETER 1922, KUNZE 1994). The shape of the corpusculum becomes visible with the start of corona differentiation, while the caudicles and pollinia are overgrown by wing-like anther margins until shortly before the opening of the flower (Fig. 9E, F). The completion of the caudicles, and thus the final attachment of the pollinia happens as a last step (cf. KUNZE 1994).

Pollinaria. The pollinia are flattened ovoid to D-shaped, 270–550 µm long, 200–350 µm in diameter and yellow to brown-orange in colour. Size and sometimes colour are taxon-specific. The pellucid margins (germination crests), a synapomorphy of the tribe *Stapelieae* (BRUYNS & FORSTER 1991), are comparatively massive (Fig. 10A, B) and typically lifted up by translucent, ca. 100 µm high rims (Fig. 10C). It is characteristic of the genus that the pellucid margins are affixed along the upper, outer half of each pollinia in subapical to lateral position (Fig. 10A). Rarely, the rims of the margins are fully parallel (*D. immaculata*); in most cases they are broadened in the middle, resulting in an elliptic outline in top view (*D. angustiloba*, *D. sulcata*, Fig. 10A, B). The number of pollen cells per pollinium amounts to 150–250. Starting from linear microspore tetrads, the three-celled pollen grains (cf. DANNENBAUM & SCHILL 1991) organize themselves in rows of four to six grains directed towards the germination margin. The caudicles are affixed to the pollinia in a terminal to subterminal position (cf. SCHILL & JÄKEL 1978). Caudicles are divided into a short stalk and a disc- to a bowl-like apical part, which ventrally corresponds to the pollinia basis and connects with it (Fig. 10D, E). The stalk extends basally into the massive caudicular wings, which are laterally affixed to the broadly elliptical corpusculum for the basal third. The function of these wings is hypothetical. Probably their adnation to the stylar head builds up a supporting resistance against the pull of the fly labellum removing the pollinarium, so that the cleft of the corpusculum becomes firmly attached to the bristles of the labellum (cf. MEVE & LIEDE 1994).

After the removal of the pollinaria from the stylar head the pollinia tilt downward, so that the germination margin comes into an upward position while the pollinarium is fixed at the fly's proboscis, enabling its insertion into the guide rails (Fig. 10F). Shrinking movements of the drying caudiculae, widespread in many stapeliad genera (cf. JAHNKE 1989), are not found or are insignificant.

Fruit development. After shedding of the flower, the gynoecium rests at first without further growth. The sepals, which are still attached to the pedicel, wrap themselves protectively around the carpels and cover them almost completely. In some species the fruit stalks curve to an erect position (Fig. 5D). Sometimes the pedicel swells and becomes secondarily succulent. The fruit rest takes at least 80 days (greenhouse observations in Münster), normally it takes 100–180 days (Fig. 11A). In extreme cases, fruit rest can continue for several years. The longest fruit rest observed in *Duvalia* was almost four years. The degree of succulence of the fruit stalks is usually correlated with the duration of dormancy. The thicker the fruit stalk, the longer the resting phase of the undeveloped fruits.

The fruit dormancy obligatory for the stapeliads has been studied in a few cases. VOLK (1944/1949, 1951) established that in *Orbeopsis lutea* and *Lavrania* spp. fruit development only starts in the vegetative period following pollination. BALLY & al. (1975) noted that in *Pseudolithos*, fruit dormancy took as long as 18

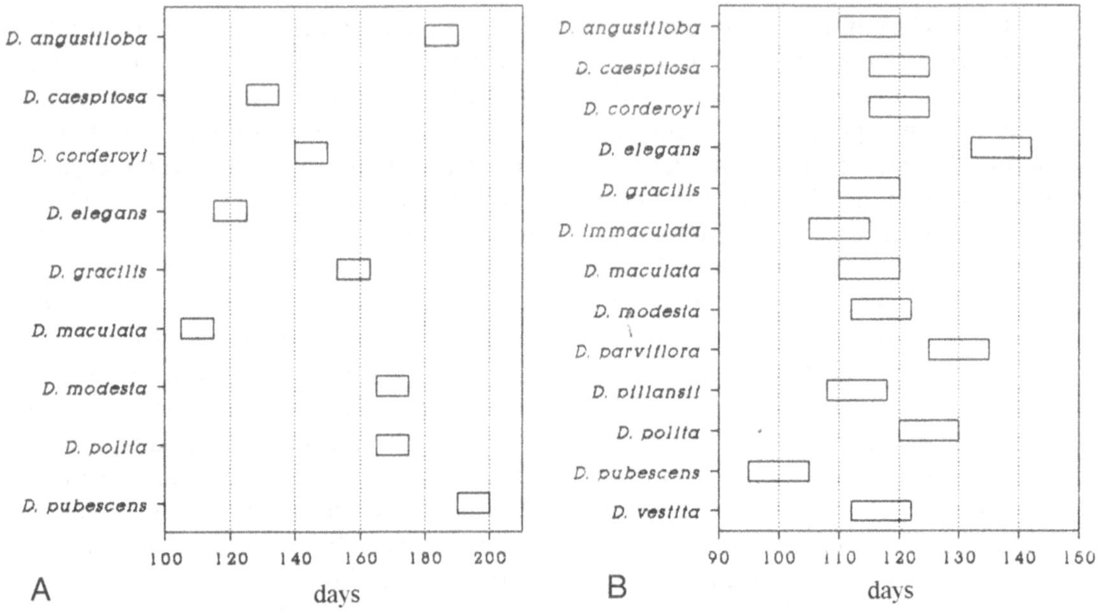

Fig. 11. Duration of fruiting in *Duvalia* sect. *Duvalia*. *A* Rest phase between fertilization and growth of fruits in days; *B* growth time plus time of ripeness in days. Observations were made in the greenhouse, Münster Botanic Garden, Germany

months. And in a plant of *Orbeopsis lutea* (N.E. Br.) Leach with several fruit initials on the same rachis, a successive maturing of the follicles over three years was observed in Münster. Barad (1990) observed the same phenomenon in *Caralluma tuberculata* N.E. Br.

The retardation of fruit development poses the question as to the regulation mechanisms. The growth inhibition is of postzygotic nature as shown by microscopic sections of fruit initials and embryo sacs. Therefore, a hormonal regulation of embryo development can be postulated. Further investigations point to a genetically rather than an inductively regulated initiation of growth. Following greenhouse observations, water availability or nutritional factors can be excluded as inductive environmental signals. It is impossible to break dormancy by external factors, such as intensive watering of the plants, shading or extra light. Strong, healthy plants produce fruits after exactly the same period of dormancy as strongly dehydrated or even dying ones. However, the relatively uniform induction of fruit growth in *Duvalia* in February/March (in Münster) points to an inductive influence of the light factor. Traversing a certain day length (in the greenhouse about 10 weeks after the winter solstice) is probably the decisive signal to break the inhibition. In the southern hemisphere, therefore, fruit growth should start around the end of August. Personal observations in the field confirm this assumption; developing stapeliad fruits are found in the Cape Provinces in spring and early summer, independent of the location, whether in the winter rainfall or the transitional rainfall area. Adding the developmental period of 12 weeks, seeds will be shed during the dry season in summer. Dry seed, seed hairs and reduced resistance of the thin vegetation promise best dispersal conditions during this time of year. After the usually quick germination (see below) the seedlings then enjoy the full period of moderate growing conditions.

Generative cycles of such rhythmicity can be understood as adaptation to the climatic periodicity in semi-arid habitats. Similar rhythms have been found in the stem-succulent *Sarcostemma* R. Br. (Howie 1991).

Survival rate of fruit initials. The dormant fruit initials are subject to a highly variable mortality rate. Successfully fertilized carpels sometimes die after several weeks or even months.

Beyond the natural mortality rate fruits are often eaten by parasitic insects, especially *Tephritidae*. Their larvae frequently destroy all seeds in maturing fruits (Meve 1995).

Fruit growth. When dormancy is definitely broken, the developing follicles grow continuously. The growth period is about 6–8 weeks. A maturing period follows, which takes about the same time or longer. On average, ca. 115 days pass between the onset of growth and dehiscence of the follicles (Fig. 11B). There may be slight species-specific deviations from this schedule, e.g., in *D. elegans*, *D. pubescens*. The short growth period of *D. pubescens* may be an adaptation to the extremely arid habitats in southern Namibia. Water requirements as well as the risk of being eaten are minimized by short fruit-growing and -maturing periods.

Fruit morphology. After fertilization usually both carpels develop into slender, fusiform, ventrally slightly flattened follicles of 6–18 cm length. They have a stalk which can swell from ca. 2 to 5 mm in diameter after the end of anthesis. The fruits are erect with an angle of 40–90° between the two follicles, and each follicle contains 40–120 seeds. The glabrous pericarp of fully developed follicles is thin-walled of ca. 500–1000 µm in diameter. The lacticifers criss-crossing the tissue are unbranched in contrast to those of the shoots. Occasionally rod-shaped cystoliths are found in the mesocarp. During maturation, the cell walls of the endocarp are thickened and lignified providing the necessary stiffness to the erect fruits, an effect enhanced by the ladder-like arrangement of cells. At maturity, the pericarp dries up and the resulting tension effects the opening of the fruit along the ventral suture (opening fruit).

Seed morphology. As usual in *Asclepiadaceae*, the seeds are relatively simple, but possess specific adaptations to anemochorous dispersal (Fig. 12). They are strongly flattened, ellipsoidal, 3–8 mm long and 2–5 mm broad. The size of the seeds is usually correlated with the flower size of the species. The light brown to black testa is laterally extended into a 0.2–1.2 mm broad seed wing (Figs. 12, 13). In micropylar position, directly in front of the hilum, the testa produces a bunch of hairs (coma) of 10–20 mm length (Fig. 12B). The seed wing is formed by the connate epidermes of the dorsal and ventral seed surface. Despite some infraspecific variability the morphology of the seed wing can be used to differentiate between some taxa: *D. gracilis* (Fig. 12F) versus *D. maculata* (Fig. 12E), or *D. elegans* (similar to *D. pubescens*, Fig. 12D) versus *D. vestita* (Fig. 12C). The cell pattern of the dorsal testa is homogeneous with isodiametrical cells of relatively flat and smooth periclinal walls. In contrast, the ventral side of the wing epidermis (Fig. 13A) is puffed up to a richly pitted palisade tissue (see Sylla & Albers 1989). Pits occur to a variable extent (Fig. 13B) or are absent. In mature seeds investigated under the SEM, the dehydrated palisade tissue reflects the

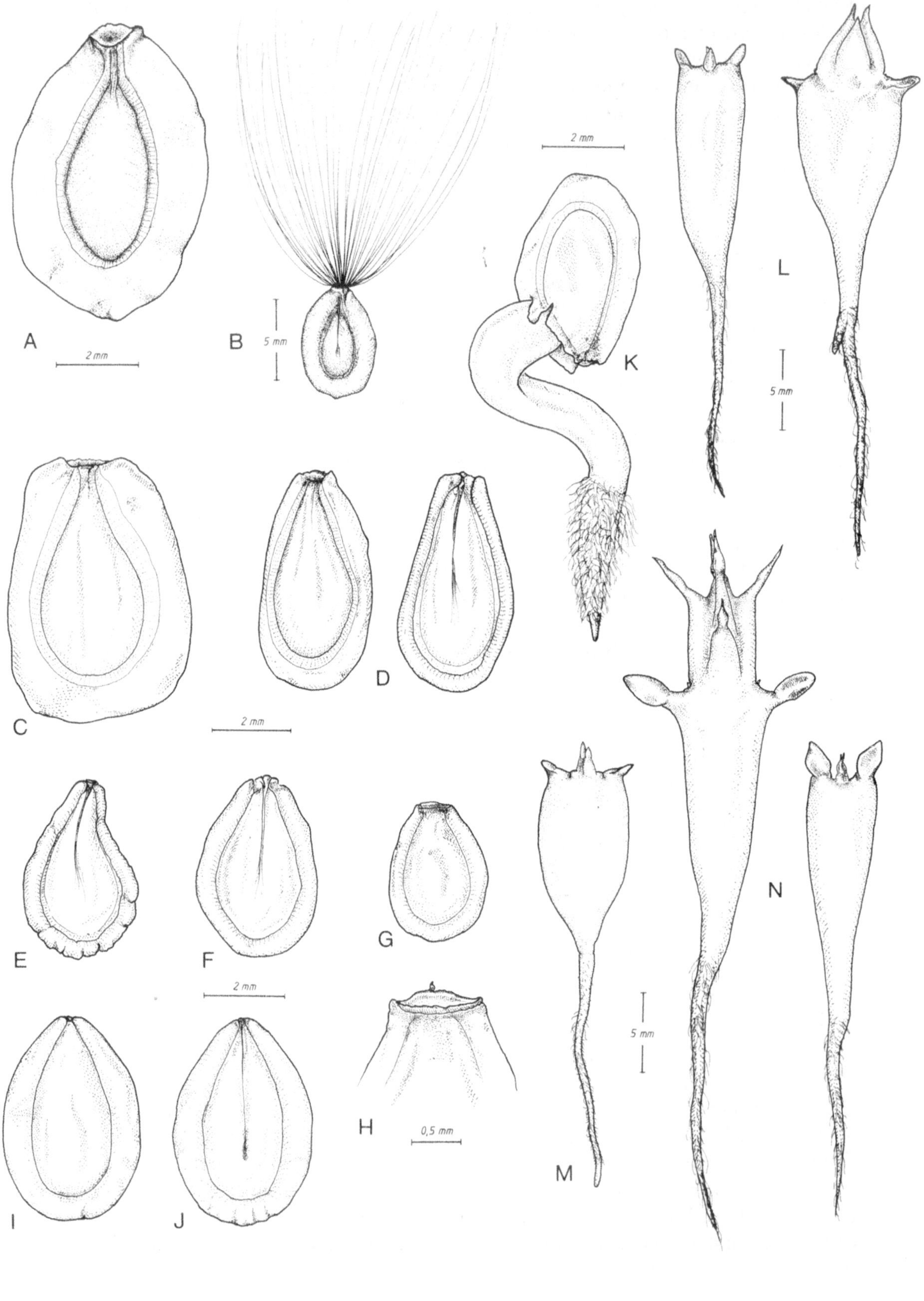
A
2 mm
B
5 mm
2 mm
K
L
5 mm
C
D
2 mm
E
F
G
2 mm
H
0,5 mm
I
J
M
5 mm
N

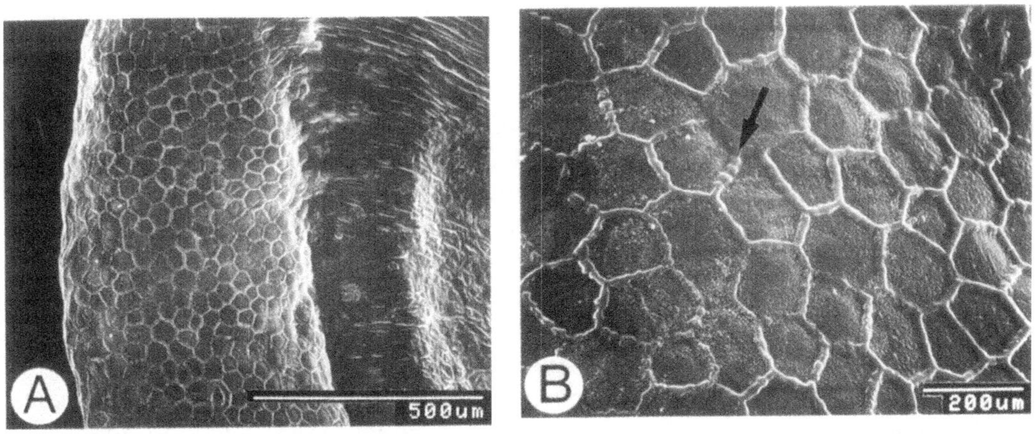

Fig. 13. *A*, *B*. Ventral side of seed wing in *D. gracilis* (Meve 346); the arrow points to a pitted anticlinal epidermal wall

distribution of the pits. The thick anticlinal walls (cell borders) are raised to bridge-like sculptures, and, if these walls are strongly pitted, the raised sculptures appear moniliform (Fig. 13B). The periclinal walls never possess pits. It seems that pitted periclinals, which are marked by round to elliptic depressed surfaces, represent a rare feature in *Asclepiadaceae*. So far, they have been demonstrated only in the stapeliad *Lavrania* Plowes (Bruyns 1993: 172) and the marsdeniad *Hoya* R. Br. (Rahman & Wilcock 1991a).

The mature seeds are almost completely filled by the linear, flat embryo, which is surrounded only by a degenerating, 1–2 layered endosperm. The radicle and the voluminous oil-rich endosperm are straight. The plumule is covered by the cotyledons. Forced by the flat shape of the seed, the cotyledons show an obliquely diagonal position, which is retained in the seedling (Fig. 12L, M).

Germination and seedlings. Stapeliad seeds quickly germinate after moistening, normally within 2–10 days. Water imbibition is accomplished by the remainders of bundles of the suture (Sylla 1986). After ca. 2 to 3 days the growing radicle breaks through the testa on the dorsal surface of the seed in the region between the seed body and the seed wing (Fig. 12K). Germination under partial rupture of the testa only takes place in this position close to the micropyle. The

---

Fig. 12. Seeds and seedlings of *Duvalia*. *A–J* Seeds: *A* dorsal side, *B* seed with coma (*D. polita*, Meve 470); *C* dorsal side (*D. vestita*, Meve 397); *D* left dorsal side, right ventral side (*D. pubescens*, Meve 218); *E* ventral side (*D. maculata*, Meve 457); *F* ventral side (*D. gracilis*, Meve *345)*; *G* dorsal side; *H* micropylic region (*G*, *H* from *D. parviflora*, Meve & Kusch sub K 1562; *I* dorsal and *J* ventral side (*D. eilensis*, from Lavranos, Barad & Linden 24952); *K* germinating seed with radicle bursting through testa *(D. caespitosa*, Meve 388). *L – N* Seedlings: *L* left one week old, right ca. weeks old (*D. pubescens*, Albers & Meve 33); *M* ca. two weeks old (*D. caespitosa*, s. n.); *N* left ca. eight weeks old, right two weeks old (*D. polita,* Albers & al. 528)

cells forming the micropyle itself are entirely used for seed hair formation (Fig. 12H). After loss of the coma a hard, slightly sunken scar remains, so that, in contrast to most other seed plants, this region is unsuitable for the emergence of the radicle. Depending on temperature, the radicle and the hypocotyl elongate within the next 24–36 hours until the seedling unwraps from the testa. After successful fixation in the substrate, the seedling straightens up vertically above the substrate and expands its cotyledons (epigeal germination) (Fig. 12M). The seedling remains in this stage for some days or weeks, even months under unfavorable conditions, before the primary and subsequent leaves appear (Fig. 12L, N). With exception of the radicle the whole seedling is succulent; stem succulence is thus fixed as early as in the seedling (or embryo, see SYLLA & ALBERS 1989).

The rounded spathulate cotyledons always lack stipules. Their size is species-specific and correlated with the size of the subsequent leaves. *Duvalia polita* with the largest leaf rudiments in the genus also possesses the largest cotyledons (Fig. 12N).

Chromosomes and ploidy levels

Chromosome numbers. *Asclepiadaceae* are characterized by the base number x = 11. Very few deviations (x = 10, x = 9) are known (ALBERS & al. 1993), and one or two B-Chromosomes occasionally have been observed (ALBERS & MEVE 1991). Chromosomal information has been presented by ALBERS (1983) and ALBERS & MEVE (1991). Counts for some *Duvalia* species have been previously published by REESE (1971), REESE & KRESSEL (1967) and ALBERS (1976, 1977).

Duvalia is karyologically diverse with di-, tetra- and hexaploid species, though diploid chromosome sets of 2n = 22 are still to be found in the majority of the species (Table 4, Fig. 14). All four African taxa of sect. *Arabica* have 2n = 22, and only in the Arabian subspecies of *D. sulcata* a doubling to tetraploid genomes of 2n = 44 took place. In sect. *Duvalia* diploidy diminishes significantly in favour of polyploidy.

Polyploidy occurs to an extent of 50% in sect. *Duvalia*, respectively 33% in the whole genus. Normally, variation in stapeliad genome size is comparatively low, only about 10% of the taxa have been found to be polyploid (ALBERS & MEVE 1991). Only the genus *Orbea* shows a higher percentage of polyploids (almost 44%), followed by *Orbeopsis* (20%), *Echidnopsis* (17%) and *Huernia* with 12% (ALBERS & MEVE, unpubl.).

In *Duvalia*, as well as in the similarly disjunct genus *Orbea* (cf. LEACH 1978), it is remarkable that the polyploid species are restricted to southern Africa. This might be a hint as to the causal link of changing environmental conditions and a high level of polyploidy, because the dry areas of East Africa and Arabia are known to be very old and stable.

The southern African polyploid series. In sect. *Duvalia* every second species is a polyploid with a tetra- or hexaploid genome (Table 4, Fig. 15). The distribution areas of almost all diploid species are considerably smaller than those of the polyploid ones. Moreover, the areas of the polyploids are usually situated in peripheral regions (with *D. maculata* as a diploid exception). The hexaploid *D. corderoyi* and *D. immaculata*, as well as the tetraploid *D. pubescens*

Table 4. Chromosome numbers in all *Duvalia* taxa. (Numbers in brackets indicate single deviating counts)

Taxon	Chromosome number (2n)	Reference
Duvalia sect. *Duvalia*		
D. angustiloba	22	REESE (1971)
D. elegans	22	ALBERS (1976)
D. gracilis	22	this paper
D. maculata	22	this paper
D. modesta	22	this paper
D. parviflora	22	ALBERS (1977)
D. pillansii	22	ALBERS (1977)
D. polita	44 (22)	REESE & KRESSEL (1967)
D. caespitosa var. *caespitosa*	44	REESE & KRESSEL (1967)
D. caespitosa var. *compacta*	44	REESE & KRESSEL (1967)
D. pubescens	44	this paper
D. vestita	44	this paper
D. corderoyi	66	REESE & KRESSEL (1967)
D. immaculata	66	ALBERS (1977)
Duvalia sect. *Arabica*		
D. eilensis	22	this paper
D. galgallensis	22	this paper
D. sulcata subsp. *sulcata*, Africa	22	this paper
D. sulcata subsp. *sulcata*, Arabia	44	REESE (1971)
D. sulcata subsp. *seminuda*	22 (44)	REESE (1971)
D. sulcata subsp. *somalensis*	22	this paper
D. velutina	22	this paper

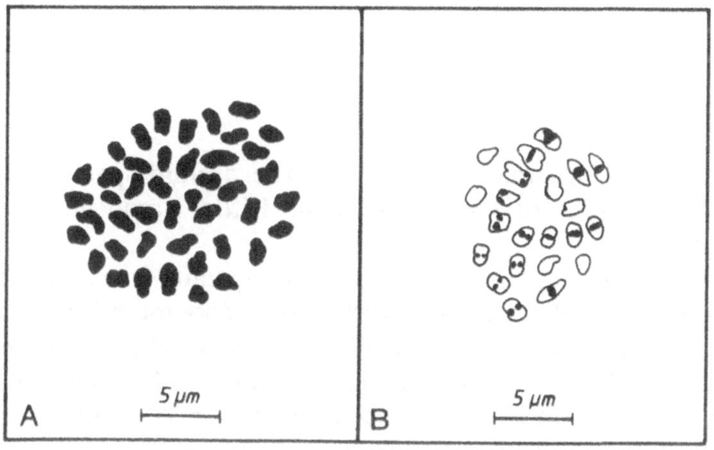

Fig. 14. Somatic metaphase plates from root tips of *Duvalia* species. *A D. pubescens*, MEVE 232, 2n = 44 (carmine); *B D. angustiloba*, MEVE 324, 2n = 22 (Giemsa C-banding)

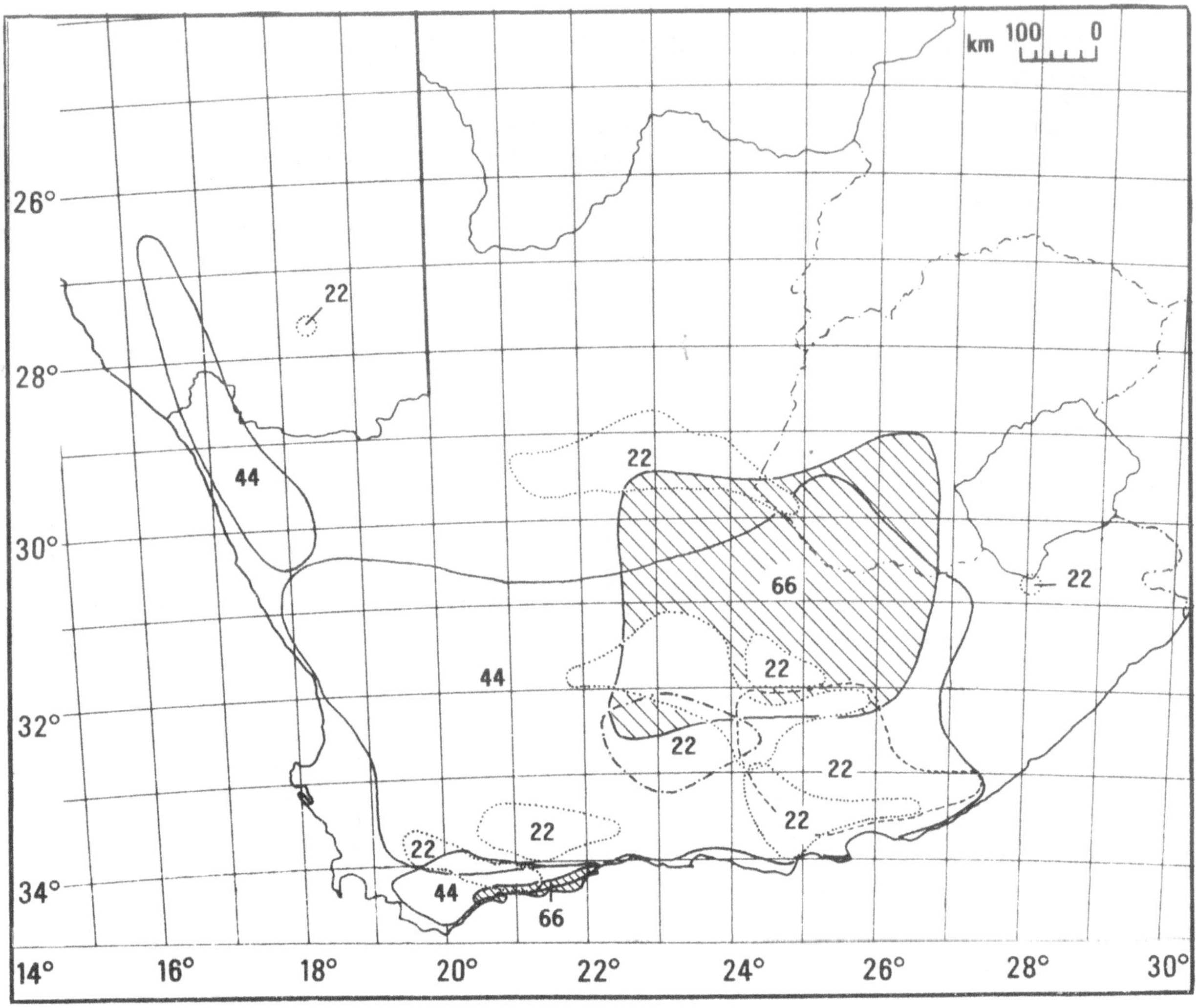

Fig. 15. Species distribution areas and ploidy levels of *Duvalia* in southern Africa (for individual chromosome numbers and distribution areas of the taxa involved see Table 4 and Taxonomic treatment)

and *D. vestita* show a peripheral distribution (Fig. 15). The diploid species are concentrated in the diversity centres of the genus in the Little Karoo and the karroid areas of the Eastern Cape.

D. maculata has a rather disjunct distribution with two main areas in the Eastern and Northern Cape and a small area in the Karasberge of southern Namibia (Fig. 20). Because long distance dispersal is unlikely, such scattered populations obviously represent relict localities. Comparatively unstable and changeable environmental conditions which have probably been in existence in southern Africa for the last 2 million years (Goldblatt 1978, Van Zinderen Bakker 1978) might be responsible for this area fragmentation. These unstable conditions, however, also should have caused the development of quickly changing, and thus newly accessible habitats. New, unoccupied habitats are an important precondition for newly evolved "neopolyploids" to establish themselves apart away from the competing diploid parents, which are normally better adapted to their place of origin (EHRENDORFER 1980). "Because of the fast effective hybridogenic variability and the resulting better capacities for adaptation and expansion, they manage to escape the competition of their parents or to compete successfully against their parents" (translated from EHRENDORFER 1980, cf. also EHRENDORFER 1962). Polyploid series or complexes within higher plants often show astonishing parallels. At their starting points we find more or less isolated diploid species and, maybe, individual neopolyploid elements. Development then continues often with polyploids of different levels until an optimum of diversity is reached. Finally, the diploids, their distribution area and the whole diversity decreases, and at the end scattered polyploids remain (EHRENDORFER 1980). Section *Duvalia* obviously represents a young to medium-aged polyploid series, because a considerable number of diploid species still exists, although with rather restricted distribution areas. On the other hand, polyploid species are not restricted to newly conquered karroid habitats. The hexaploid *D. immaculata*, geographically isolated in the south between the Langkloof – and Langeberg escarpments and the Ocean, has even switched from marginal Karoo to Coastal Fynbos habitats.

With the tetraploid *D. vestita*, which is also partly found in Fynbos (MEVE 1988), this ecological transposition should have originated in competition with the parental clones. *Duvalia vestita* exhibits a typical "strategy of escape" that is often observed in neopolyploids (cf. EHRENDORFER 1980).

Heteroploid taxa. *Duvalia polita* is a heteroploid species in which about 15% of diploid individuals occur apart from the widespread tetraploid ones. Out of a total of 23 plants investigated, the three diploid ones are clearly limited to the central eastern part of the species distribution area (Zimbabwe, RSA: Transvaal; Fig. 16), where they occur occasionally in predominantly tetraploid populations.

Duvalia sulcata subsp. *sulcata* is clearly separated into two cytogeographical races (cytotypes). The African populations are diploid (4 counts), whereas the Arabians are tetraploid (3 counts).

Within the normally diploid *D. sulcata* subsp. *seminuda* only one tetraploid plant (out of a total of seven plants sampled) was documented for the southwest of the Arabian Peninsula, probably representing a spontaneous cytological deviation.

The origin of polyploidy. The formation of polyploid genomes may be due to "autoploidy", "alloploidy" or "autoalloploidy" (cf. STACE 1989). Diploid *Duvalia* species normally possess rather distinct and constant morphological features which makes them easy to recognize. In contrast, the tetraploid species *D. caespitosa*, *D. pubescens* and *D. vestita* are highly variable in vegetative and floral characters. Also, they are fairly widely and often peripherally distributed. These three species may be considered as allopolyploids, typically exhibiting a wide range of variability as a consequence of recombination between parents of widely different characters (cf. EHRENDORFER 1991).

The two heteroploid species, *D. polita* and *D. sulcata*, show little variability independent of ploidy level. The lack of any phenotypic differences or habitat preferences between the two cytotypes gives evidence for an autopolyploid origin of the tetraploid genomes.

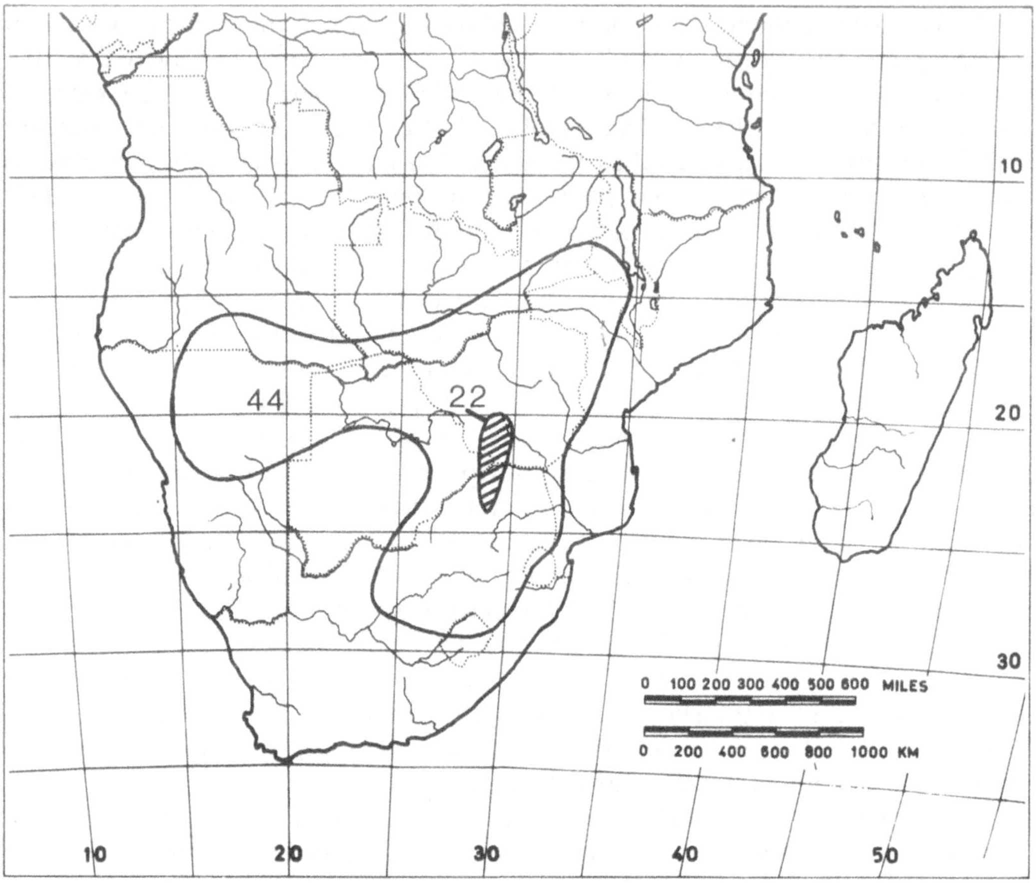

Fig. 16. Distribution of the two cytotypes in *Duvalia polita* with 2n = 22 (hatched) and 2n = 44 (bare)

The sporadic occurrence of diploids within *D. polita* and their geographical restriction to the centre of distribution in contrast to the widely radiating tetraploid cytotype is characteristic of a polyploid complex (cf. LEWIS 1980). In the case of *D. polita* a still incomplete polyploidisation of the species can be assumed. The distribution area of the diploids should thus indicate the centre of origin of the species.

The hexaploid species *D. corderoyi* and especially *D. immaculata* inhabit peripheral parts of the distribution area of sect. *Duvalia*. Both show low morphological variability. Presumably they originated from hybrids between di- and tetraploid parental species with subsequent doubling of the then triploid chromosome set. This autoallopolyploid origin (cf. STACE 1989) runs parallel to a phenotypical stabilisation.

Chromosome morphology. Chromosomes usually measure 1.0 to 1.9 µm in length (Fig. 14). This relatively small chromosome size limits the application of elaborate investigation methods (e.g., Giemsa C-banding, SCHWARZACHER & al. 1980). Apart from technical problems, this method has been shown to be of little benefit in the stapeliads because of the lack of assessable variation in the stainable heterochromatic bands of the chromosomes (ALBERS 1983). *Duvalia* is characterized by 16 meta – to submetacentric chromosomes, 14 of which are banded in centromeric and two in telomeric position (Fig. 14B, after Giemsa treatment), while six chromosomes remain unstained lacking heterochromatin. This pattern does not deviate much from those of other stapeliads, e.g., *Quaqua incarnata*, in which all 16 chromosomes with heterochromatin were stained in centromeric position (ALBERS 1983).

Nevertheless, in comparison to many other *Asclepiadaceae*, in which the chromosomes often only reach half the size of stapeliad chromosomes (e.g., in the tribe *Asclepiadeae*, ALBERS & al. 1993), the chromosomes in the *Stapelieae* often are relatively large, and possibly represent the largest ones in the whole family (MEVE & ALBERS, unpubl. data on all tribes of *Asclepiadaceae* s. str.).

Pollination

As most stapeliads, *Duvalia* exhibits a sapromyiophilous pollination syndrome (MEVE & LIEDE 1994). The flowers emit stenches of excrements and carrion for olfactory attraction. Various macro-diptera were observed as pollinators in the wild as well as in the greenhouse. Most dominant are flesh flies and house flies (*Calliphora, Lucilia, Sarcophaga, Musca*). Occasionally, other flies (e.g., *Tachinidae*) can be seen on *Duvalia* flowers. Pollination is initiated when the flies, while probing or licking nectar, approach a guide rail entrance with their proboscis, so that hairs of the proboscis get caught in the guide rail. When the flies try to escape, the hairs are led upwards in the guide rails until they finally lodge in the cleft of the corpusculum, and thus release the whole pollinarium (Fig. 10F; for details see MEVE & LIEDE 1994).

Breeding system

Inter- and infrageneric hybrids of wild origin. Natural hybrids have been reported for *D. caespitosa* with *Huernia pillansii* (BRUYNS 1981b, ALBERS & MEVE 1991), *Duvalia* spec. × *Huernia clavigera* (JACQ.) HAW. (LEACH 1988), *D. corderoyi* × *Huernia humilis* (P. BRUYNS, pers. comm.). An intergeneric hybridisation of *D. caespitosa* with *Piaranthus* spec. led to the description of x *Duvaliaranthus albostriatus* BRUYNS (1976). In the present work the general hybridisation potential of *Duvalia caespitosa* with *Piaranthus*, here *P. comptus*, was proven by hand-pollination in the greenhouse. The result strongly supports the parental species assumed by BRUYNS (1976): the flowers of the hybrid created do not show any significant differences to those of x *Duvaliaranthus albostriatus*. Interspecific hybrids seem to be rare in *Duvalia*. Plants assumed to have resulted from natural hybridisation between *D. parviflora* and *D. caespitosa* (sect. *Duvalia*) and between *D. sulcata* subsp. *seminuda* and *D. velutina* (sect. *Arabica*) have been collected (see taxonomic treatment).

Artificial pollination. Artificial pollination and successful fertilization is possible within and between many stapeliad genera (for a survey see BARAD 1990). In *Duvalia* a considerable number of taxa can be crossed artificially. This is documented by a broad crossing programm (Table. 5). A cross has been considered successful when well-developed seeds were set. Germination experiments

Table 5. Results of artificial pollination in *Duvalia*, based on Bayer & Harold (1987) and own investigations. (*f* fertile; *s* sterile; *i* mechanically isolated; – no result)

	Duvalia taxa																	
	1.	2.	3.	4.	5.	6.	7.	8.	9.	10.	11.	12.	13.	14.	15.	16.	17.	self
Sect. *Duvalia*, 2n = 22																		
1. *D. angustiloba*		fs	fs	fs	fs	fs	s	–	fs	f	–	s	s	i	i	i	s	s
2. *D. elegans*	fs		–	f	f	f	f	fs	fs	f	fs	fs	s	fs	–	s	–	s
3. *D. gracilis*	fs	–		fs	fs	–	–	–	–	–	s	–	–	–	–	–	–	s
4. *D. maculata*	fs	fs	fs		fs	s	s	fs	s	s	–	–	s	f	fs	–	–	s
5. *D. modesta*	fs	fs	–	f		fs	f	f	fs	fs	–	fs	fs	–	s	–	fs	fs
6. *D. parviflora*	fs	s	–	fs	f		f	s	–	f	s	–	fs	fs	i	i	i	s
7. *D. pillansii*	f	f	–	f	fs	s		f	s	fs	s	–	fs	s	–	–	–	s
2n = 44																		
8. *D. caespitosa* var. *caespitosa*	f	f	fs	fs	fs	f	s		f	fs	fs	fs	fs	f	s	s	–	f
9. *D. caespitosa* var. *compacta*	s	f	–	s	fs	s	s	f		–	fs	s	–	–	–	–	–	fs
10. *D. polita*	–	s	s	s	f	s	s	fs	–		–	–	fs	f	s	s	–	fs
11. *D. pubescens*	s	s	–	–	–	f	–	fs	fs	fs		fs	–	–	–	–	–	f
12. *D. vestita*	–	–	s	–	–	–	–	f	fs	–	s		–	–	–	s	s	f
2n = 66																		
13. *D. corderoyi*	s	s	–	s	s	s	s	fs	s	s	–	–		s	–	s	–	f
14. *D. immaculata*	s	s	–	s	s	s	s	s	–	s	–	–	s		–	i	i	f
Sect. *Arabica*, 2n = 22																		
15. *D. eilensis*	s	–	–	–	–	–	–	–	–	s	–	–	–	–		s	–	s
16. *D. sulcata* var. *seminuda*	–	–	s	–	–	–	–	s	fs	–	–	s	s	–	s		s	s
17. *D. velutina*	–	–	–	–	–	–	–	–	–	–	–	–	–	–	–	fs		–

showed that the F_1-generation is normally fertile. An F_2-generation has not been examined.

The crossability of the most frequent taxa of sect. *Duvalia* has been studied by BAYER & HAROLD (1987). Their almost uncommented published results have been included in the present author's own investigations (Table 5).

In many cases, the use of different individuals of the same species may lead to different results. Therefore, failure of crossing two species in a particular case does not mean that hybridization is impossible. From the large number of individuals studied some remarkable tendencies can be recognized. Diploids and tetraploids of southern Africa are widely crossable (Table 5). This is true for crosses within the same ploidy level as well as between different ploidy levels. Only *D. polita* is an exception, because this species is almost completely incompatible with other members of both sect. *Duvalia* and sect. *Arabica*. These results confirm the exceptional position of *D. polita* indicated by the biogeographical and morphological peculiarities of this species.

The hexaploids *D. corderoyi* and *D. immaculata* must be understood as reproductively isolated. Hand-pollinated flowers can never be fertilized with extraneous pollen and they are poor pollen donors for successful crosses (Table 5).

The obvious contrast between self-sterile diploids and self-fertile polyploids demands attention and needs to be discussed in connection with the expanded distribution of polyploid *Duvalia* species in southern Africa. In the vast areas of the Karoo, potential crossing partners are widely spaced and may easily be missed. Self fertility allows for more independence in space and thus increases the possibilities for area expansion by a potentially much higher seed production on the one hand and guaranteed sexual reproduction even in isolated plants on the other hand. Even though 90% of the stapeliads are diploids (ALBERS & MEVE 1991), and these diploids are usually self-sterile, e.g., *Piaranthus* (MEVE 1994), the influence of the breeding system for the radiation of stapeliad species, however, has to be considered separately in every genus.

Only few crosses between members of different sections have been carried out successfully. G. BARAD (pers. comm.) has managed to hybridize *D. sulcata* subsp. *seminuda* with *D. caespitosa* var. *compacta*. Restricted crossability points to a long independent and divergent development of both groups. *Duvalia eilensis* proved to be completely isolated gametophytically within sect. *Arabica*. *Duvalia sulcata* and *D. velutina* can be crossed relatively easily (see above).

Isolation mechanisms. Mechanical isolation mechanisms. In the field, mechanical incompatibilities may serve as successful prezygotic isolation mechanisms. Pellucid margins of the pollinia and guide rails often do not fit together due to different sizes. For example, the large pollinia of *D. corderoyi* cannot be introduced (or only by manual force) into the small guide rails of *D. angustiloba* and *D. maculata*. Or, in contrast, the pellucid margins can be too small, sliding through guide rails which are too large, for example, *D. caespitosa* var. *compacta* or *D. modesta* pollinia in *D. caespitosa* s. str. gynostegia. In the greenhouse, such barriers can be overcome by skilful manipulation.

Geographical and ecological isolation mechanisms. The astonishingly easy and free crossability of many species contrasts with the few known hybrids of wild origin. Why is this so?

Geographical isolation and vicariance play a major role in separating the species. Despite almost identical habitat preferences common populations of two (or more) *Duvalia* species of the same ploidy level are not known or are extremely rare. In contrast, mixed populations of a diploid species, and, e.g., the tetraploid *D. caespitosa*, do exist frequently. Interestingly, the other tetraploid species (those being in danger of hybridization) are allopatrically distributed.

Populations of *D. corderoyi* (2n = 66) mixed with *D. maculata* (2n = 22) or *D. caespitosa* (2n = 44) can also be found. In general, species with genomes of the same size are usually allopatrically or parapatrically (only rarely partially sympatrically) distributed. The genome of every species thus influences the present distribution of the genus. The (cryptic) basis for geographical isolation is in many cases gametophytical isolation.

In sect. *Arabica* the geographical isolation of many species is accompanied by an ecological one, so that there is a clear vicariance: *D. velutina* is a lowland species and mostly isolated from *D. sulcata*, a species of medium and high altitudes.

Flavonoids

Flavonoids are of limited occurrence in *Asclepiadaceae*; quality and quantity of flavonoid aglycons and glycosides is much lower in this family than in most others. Nevertheless, some information is scattered over the literature, proving a widespread content of luteolin (HEGNAUER 1989), quercetin and kaempferol (RAHMAN & WILCOCK 1991b, SUBRAMANIAN & NAIR 1968). In addition to the pure flower cyanidins investigated by KNAAK & al. (1969), stems and leaves have been studied.

Duvalia stems analysed only possess traces of the flavonoid aglycons quercetin and luteolin. E. WOLLENWEBER (pers. comm.) investigated *D. caespitosa* (voucher: MEVE 252) with respect to externally accumulated flavonoids. Again, small amounts of luteolin and one undetermined flavonoid have been identified in the thin epicuticular coating.

More useful patterns were obtained by extracting flowers (fresh or deep-frozen) and investigating them for the occurrence of six different flavonoid aglycons. From these, apigenin, isorhamnetin and myricetin were always absent, while kaempferol, luteolin and quercetin could be detected in most cases (Table 6). In addition, two undetermined flavonoids were spotted. However, the systematic relevance of the results is limited. Within sect. *Duvalia* variation is low. Only the presence or absence of kaempferol shows species-specifity; for instance, it allows the separation of *D. pubescens* from the closely related *D. caespitosa* (Table 6). The absence of luteolin is correlated with polyploidy.

Section *Arabica* is extremely impoverished in flavonoids (Table 6). Even quercetin, which is always present in sect. *Duvalia*, is mostly absent. This almost complete absence of flavonoids in sect. *Arabica* can be interpreted as a derived character (KUBITZKI 1967). Investigations of flavonoids throughout the whole plant kingdom, usually from the leaves, reveal a wide-spread tendency in advanced groups to replace the synthesis of flavonols by flavones (KUBITZKI 1967, HARBORNE & TURNER 1984). In addition, derived groups often show a strong reduction of the

Table 6. Distribution of flower flavonoids in *Duvalia* taxa with chromosome numbers and collection numbers given. Flavonols: *Q* quercetin; *K* kaempferol; flavones: *L* luteolin. *UF1*, *UF2* undetermined flavonoids with Rf-values given. (++ strongly present; + present; (+) weakly present; – not detected)

Taxon	No.	Q	K	L	UF1	UF2
Sect. *Duvalia*, 2n = 22						
D. elegans	UM 393	+	–	–	0.10	0.69
D. modesta	UM 328	+	–	(+)	–	0.61
D. pillansii	K 1505	(+)	–	(+)	0.09	0.62
D. parviflora	K 1562	++	+	(+)	0.12	–
D. angustiloba	UM 319	++	(+)	–	0.11	0.66
D. gracilis	UM 346	+	(+)	(+)	0.12	0.65
D. maculata	UM 457	+	–	(+)	0.11	0.66
2n = 44						
D. caespitosa var. *caespitosa*	UM 284	+	–	–	0.10	0.65
D. caespitosa var. *caespitosa*	UM 418	(+)	–	–	0.09	0.62
D. caespitosa var. *compacta*	UM 147	+	–	–	–	0.66
D. caespitosa var. *compacta*	UM 314	+	–	–	0.10	0.69
D. pubescens	UM 152	+	(+)	–	–	0.65
D. pubescens	UM 165	++	(+)	–	–	0.63
D. vestita	UM 397	+	–	(+)	0.11	0.65
D. polita	UM 470	+	+	–	–	–
2n = 66						
D. corderoyi	UM 351	(+)	(+)	–	–	0.67
D. immaculata	UM 399	+	–	(+)	–	0.68
Sect. *Arabica*, 2n = 22						
D. eilensis	L24952	(+)	–	–	–	–
D. sulcata subsp. *sulcata*	H21507	–	–	–	–	–
D. sulcata subsp. *seminuda*	FN1373	–	–	–	0.20	–

flavonoid spectrum up to the complete loss of the capacity to synthesize flavonoids (Kubitzki 1967, Mabry & al. 1970).

Phylogenetic and biogeographic considerations

"Polyploidy is a very conspicuous phenomenon in higher plants. Its study can reveal a great deal of information about the history of some groups, because polyploidisation is one of the few processes of evolution of which the direction can be determined without fossil evidence" (Stebbins 1980). Under this aspect the polyploid series within sect. *Duvalia*, and the distributional pattern of tetraploids in sect. *Arabica*, plays a central role in the discussion of phylogeny and biogeography. It reveals the differentiation in space and time (Fig. 17) and sheds light on the origin of sect. *Duvalia* in southern Africa.

***Duvalia* sect. *Duvalia*.** Today, we consider *D. polita* as a polyploid species. Regarding its relict diploidy, its prevalent polyploidy is almost certainly of autopolyploid nature. Apart from the formation of stolons, its morphological,

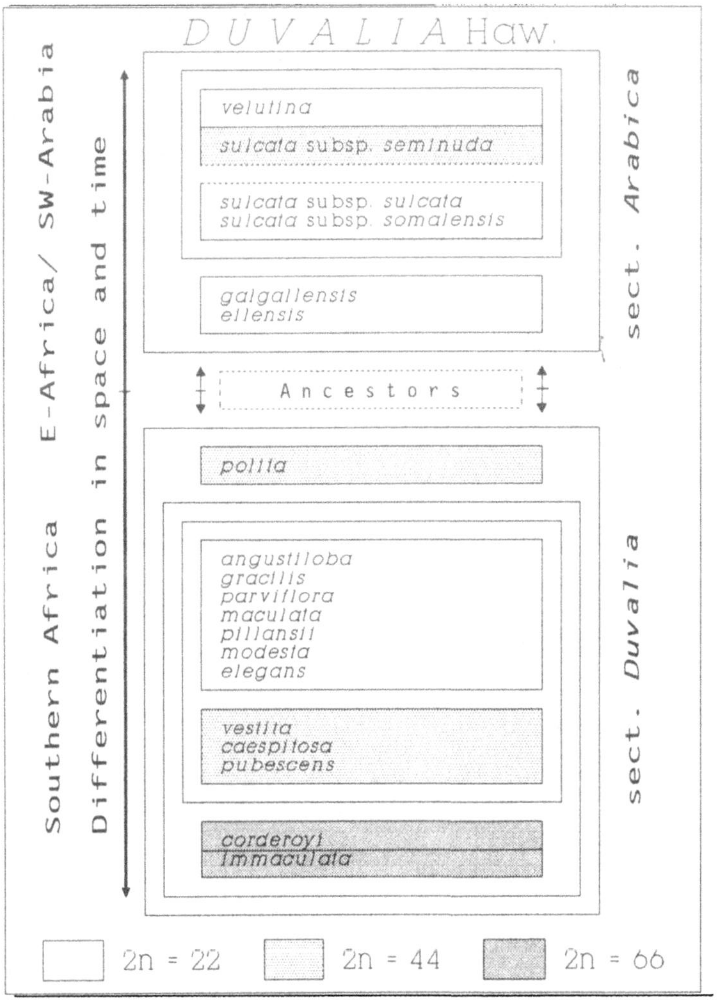

Fig. 17. *Duvalia* in space and time: a hypothetical history of evolution and migration. (The time factor is expressed by successive marginal placing of the presumably younger species; phylogentic relationships are expressed by integrating most closely related species as well as groups within boxes. See also text)

genetical and distribution characters (element of the Sudano-Zambesian Region) indicate an isolated and basal position in sect. *Duvalia*. Therefore, *Duvalia polita* should not be linked directly with the Cape species (Fig. 17). *D. polita* has probably evolved from ancestral species in East Africa then moved south into Zimbabwe and the Transvaal (where we find the diploids today, Fig. 16) and the conquest of more western habitats as far as Namibia is of secondary nature. The ancestral populations of the modern Cape Duvalias might have migrated along a similar path, parallel to the East African mountain chains, into the Karoo. Their descendants are the diploid Cape taxa, the differentiation of which is most likely due to geographical isolation. The area of highest species density in the southeastern Cape (Graaff Reinet: Grid square 3224), with as many as four diploid taxa (Fig. 2), has to be understood as a secondary centre of distribution for the genus. Starting from this centre, the differentiation of polyploid taxa has taken place during a third burst of evolution. Polyploidisation has characteristically been linked with an area expansion and habitats have been conquered which represent the extreme limit of this type of life form. The tetraploid *D. vestita* and the hexaploid *D. immaculata* finally have undergone a habitat change into the Cape plant communities, which presumably represent the most recent event in the phylogeny of Cape *Duvalia*. Genetical (gametic) incompatibilities and vicariance maintain the present distribution pattern of the Cape species (see above).

Considering the disjunct areas of *D. maculata* (and probably also *D. elegans*, – see "Taxonomic treatment") the idea of an unsteady history of such taxa emerges. Many diploid species of the Cape Provinces might at the meantime have occupied much larger areas, which have been shifting consistently to result in the relict areas we know today. The rarity of some (diploid) species (*D. gracilis, D. parviflora, D. pillansii*) supports this idea, which is paralleled by the biogeography of many other South African taxa in a number of families (e.g., Bayer 1984a,b; Jürgens 1990).

The by no means insignificant changes of climatic conditions in southern Africa during the most recent global history can be identified as the main force behind such area shifts. The origin of the Benguela current (in connection with the ice-formation in Antarctica) and the formation of an arid area with winter rainfall in the Western Cape dates back to ± 20 million years (Raven & Axelrod 1974). It could be estimated that diploid *Duvalia* ancestors arrived in South Africa roughly at this time, or a few million years later. Mainly under the influence of the Pleistocene in the late Quarternary humid and arid periods have oscillated in the Cape (Raven & Axelrod 1974, Goldblatt 1978, Van Zinderen Bakker 1978). These climatic shifts could have been responsible for extensive area shifts. Karroid formations may have expanded northwards beyond the 25° latitude and eastwards beyond the 25° E of Greenwich. In the interglacials semi-arid regions may have existed in areas today covered by Fynbos (Van Zinderen Bakker 1978). During the global changes about 2 million years ago, some then widespread diploid *Duvalia* taxa could have suffered serious area losses during a humid period. It thus seems reasonable to place the date for origin and spread of the polyploid species into a dry period before or during the early Quaternary coinciding with the rapid occurrence of karroid habitats. The arrival of *D. immaculata* in the lowlands of the southern Cape Coast can only be understood in connection with these events. With the recurrence of a more humid climate, the species has presumably 'missed' the retreat into the dry areas, but has been sufficiently adjusted due to its hexaploid genome and self fertility, so that the isolation south of the coastal mountain chains did not force its extinction. In general, the few stapeliads with a Fynbos association show some common characters. South of the 34° latitude only 1–8 species per 1°-square (degree square = 10 000 km^2) are found, but all these species possess a plagiotropous growth form and are much more frequently polyploid than the species further north (40–100% polyploid species per 1°-square, in contrast to 0–12% in the remaining winter rainfall area and in the transitional zone; Meve, unpubl. data).

Biogeographical parallels of sect. *Duvalia* with other South African stapeliads can be reported mainly for *Piaranthus* (Meve 1994) and *Stapelia* (Leach 1985). Their centres of distribution are found in the Little Karoo and the adjacent karroid areas.

In connection with the development of sect. *Duvalia* in the arid and semi-arid areas of the south African subcontinent, some characters have developed that need to be interpreted mainly as adaptation to habitat conditions. Such characters are the formation of stolons in *D. immaculata, D. vestita* and *D. polita* as well as the strongly sculptured stem epidermis of *D. caespitosa* and *D. pubescens.* Characteristically these new inventions go along with polyploidy and rather peripheral distribution areas, so that there can be no doubt as to the direction of evolution.

The diploid taxa of the central Cape displaying rather primitive morphological and chemical characters can be interpreted as paleoendemics. The small size of stems and flowers, multifloral inflorescences and a little bulged corona disc can be listed as primitive characters. The study of floral flavonoids has yielded additional support for this idea because the diploid species normally possess the richer (especially quantitatively) and thus more primitive flavonoid patterns.

***Duvalia* sect. *Arabica*.** The vegetative plant body differs scarcely in sect. *Arabica*, but is distinctly different from the South Africa species. This phenomenon also occurs in other genera with similar distribution pattern, e.g., *Orbea*. Despite their separation from their South African relatives, however, none of the Arabian species has developed a real floral novelty. Floral structure thus is extremely conservative and almost independent of habitat influences, and should be given higher weight in phylogenetic considerations here. However, this interpretation of vegetative and floral character sets in the stapeliads needs careful assessment for each genus under consideration, because in other genera the opposite situation is more likely. Vegetatively similar groups of species are now often treated as single genera despite their considerable variation in floral morphology (e.g., *Tromotriche* (Bruyns 1995) or *Pseudolithos* (Bruyns & Meve 1995). Formerly, the species involved have been associated with different genera due to superficial floral similarities.

As compared to sect. *Duvalia*, sect. *Arabica* is regarded as derived due to its more advanced leaf morphology, the lack of stipules, its larger flowers, and by the almost complete absence of floral flavonoids. Furthermore, this investigation has shown that the two sections probably evolved along independent lines and that no section can be derived from the other (Fig. 17). The gap between the geographically closest and characteristically most primitive species of the two sections, *D. polita* and *D. eilensis*, is still too broad for direct phylogenetic interrelationships. This is possibly confirmed by the crossing experiments (see "Breeding system").

Clues to the phylogeny and the directions of migration within sect. *Arabica* can be derived from the study of the *D. sulcata* complex (Figs. 17, 47). With exception of the variable flower colour and the slightly smaller stems, the Arabian representatives do not show any significant morphological differences compared with the Sudanian ones. However, the exclusive occurrence of diploid populations of *D. sulcata* subsp. *sulcata* on the African continent and of tetraploid ones on the Arabian Peninsula is a clear hint towards a conquest of Arabia from Africa. In addition, the Arabian endemic *D. sulcata* subsp. *seminuda* spreads along the eastern border of the species area, obviously representing the youngest branch of the migration and evolutionary history of *D. sulcata*. This hypothesis of a primary centre of origin in Africa and a secondary north (east) migration finds additional support by the restriction of the narrow endemics, *D. eilensis* and *D. galgallensis*, to Africa - testifying to the high age of these African floral elements (cf. WERGER 1978).

However, the splits in the area of *D. sulcata* s. str. suggest also that the species once inhabited the whole area. During the Pliocene, about 10 million years ago, the Arabian Peninsula drifted off East Africa, opening up the Red Sea and splitting the area of *D. sulcata*. The nowadays isolated (to my knowledge) Sudanian habitats therefore need to be understood as relictual. KASSAS (1956) describes the locality of Erkovit in the Sudan as "Oasis", owing its status to special topography and geography (see above). The differentiation of *D. sulcata* subsp. *seminuda* and *D. velutina* might have begun on the Arabian Peninsula from taxa similar to diploid *D. sulcata*. *Duvalia sulcata* subsp. *seminuda* and *D. velutina* have spread parallel to the Red Sea along the climatically favourable mountain chains of Tihama and Asir, as far as 20° N, assisted by the large climatic oscillations in this region and numerous geographical niches along the different mountain chains. The presence of an impressive number of neo-endemics (DEIL 1988a) finds an explanation in the fact that the Mandab Circle (NEWTON 1980) for many millions of years has been an important migration and withdrawal area for tropical African, Mediterranean and Omanian/Indo-Malayan floral elements (DEIL 1988a, KÜRSCHNER 1986, FICI 1991).

Figure 17 summarizes the ideas on the evolutionary history and phylogeny. A primary historic centre for the genus is assumed in East Africa, with the Horn of Africa probably being the starting point for a northward and southward migration resulting in the two recent secondary centres for the genus in SW Arabia (sect. *Arabica*) and in the Cape (sect. *Duvalia*). Presumed relationships at specific level are shown by boxing closely related species, and groups, respectively (Fig. 17). Lines within boxes show unknown relationships. Species in neighbouring boxes, that are however more closely related, are placed marginally so that they occur next to each other (e.g., *D. sulcata* and *D. galgallensis* or *D. elegans* and *D. vestita*).

Relationships at generic level. Before the 20th century, the classification of the species of *Stapelieae s. str.* was largely based on coronal similarities, so that only two large genera, *Caralluma* and *Stapelia*, were recognized. The subsequent division of the *Stapelieae* into smaller genera by HAWORTH (1812), N. E. BROWN (1908), LEACH (1978, 1980, 1984) and BRUYNS (1981c, 1983) leaves this level of "ad hoc - classification" searching for more natural species groups. Even if a few monotypic genera cannot be circumvented (e.g., *Ophionella*, BRUYNS 1981c), the analysis of as many character sets as possible is indispensable in the quest to circumscribe genera as potential monophyletic units: "A combination of coronal, corolline and vegetative characters are useful in circumscribing genera" (SWARUPANANDAN 1984).

Relationship with *Piaranthus*. In vegetative characters *Duvalia* is most similar to *Piaranthus*, a small genus that almost exclusively occurs in South Africa (MEVE 1994). Here, the two genera often share habitats and compete with each other as well as with other low-growing stapeliads. *Piaranthus* flowers, however, are very different from *Duvalia* since their corolla lacks an annulus and have a different epidermal pattern (cf. MEVE 1994). With the interstaminal parts strongly reduced, the corona also shows no similarities to *Duvalia*. Therefore, their common vegetative characters seem to be the result of an adaptation to equal

Table 7. Characters shared in *Duvalia* and *Huernia*

Characters	Duvalia	Huernia
Six-angled, massive stems	*D. corderoyi*	*H. kennedyana*
Strongly reduced inflorescences	in part	in part
Corolla with annulus	all species	in part
Spathulate staminal corona lobes	in part	*H. procumbens* *H. tanganyikensis*
Disc-like C(is)	most species	*H. hallii* *H. verekerii*
Broadly winged seeds	*D. polita*	most species

ecological conditions. MEVE (1994) regarded *Piaranthus* to be closely related to *Huerniopsis* N.E. BR.

Relationship with *Huernia*. Apart from striking parallels between *Duvalia* and *Huernia* in life form and biogeography, the very similar morphological characters found in both genera suggest a close relationship. This concerns mainly inflorescence morphology and corona characters (Table 7).

Huernia was identified by R. BROWN (1810) as a homogeneous group within the *Stapelieae*, however, his circumscription of the genus lacks precision. As emphasized by BROWN (1810), lobules in each sinus of the five corolla lobes characterize many *Huernia* species (e.g., *H. campanulata* HAW., *H. volkartii* PEITSCH. ex WERDERM. & PEITSCH.); however, species with a simple corolla are known as well (e.g., *H. tanganyikensis* E.A. BRUCE & P.R.O. BALLY, Fig. 18). Furthermore, such outgrowths in the sinus can be observed in other genera as well, e.g., *Tavaresia barklyi* (TH. DYER) N. E. BR., *Stapelianthus decaryi* CHOUX or *Whitesloanea crassa* (N.E. BR.) CHIOV. This feature was also observed in individuals of *D. caespitosa* (Fig. 25D).

The genus was diagnosed by N.E. BROWN (1908) more precisely, so that a clear distinction between *Huernia* and *Duvalia* became possible. In *Huernia* the gynostegium with the interstaminal corona is situated on top of the corolla ("more or less adnate to the bottom of the corolla-tube"); in *Duvalia* the lower part of the receptaculum is extended into a basal tube. The gynostegium is thus "stipitate". This character can be regarded as the most important evolutionary novelty (autapomorphy) of the genus *Duvalia*.

Referring to these criteria, LEACH (1969) correctly transferred the three closely related species *Duvalia andreaeana* RAUH, *D. procumbens* R.A. DYER and *D. tanganyikensis* (Fig. 18A) to *Huernia*, but disregarded another generic character given by N.E. BROWN (1908) that the staminal corona in *Huernia* is "simple, and often with a slight transverse dorsal ridge at its base, but no crest, wing or dorsal horn". The three species named above, however, all possess a bipartite staminal corona with an erect, spathulate outer part as in *Duvalia*. DYER (1969) critizised the 'mistake' by LEACH and demanded a transfer back to *Duvalia*. In addition, DYER

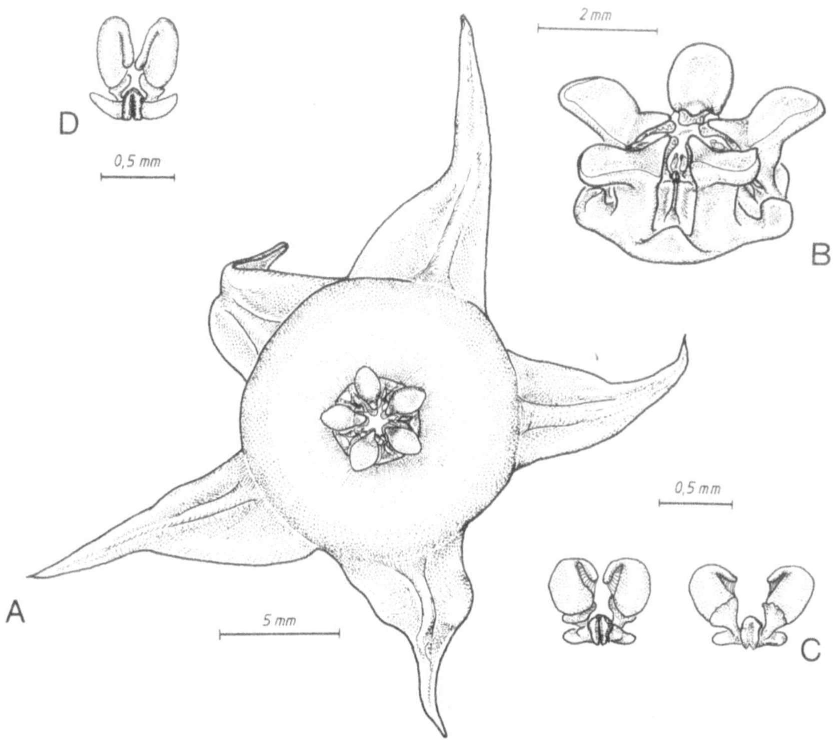

Fig. 18. *A–C Huernia tanganyikensis. A* Flower in top view; *B* corona in lateral view; *C* pollinaria, left dorsal face, right ventral face; *D. H. volkartii*, pollinarium. (*A–C* s. n. sub K 165; *D* ALBERS & al. 576)

falsely used the argument that *D. immaculata* does not possess a stipitate corona like the three transferred species. This, however, is incorrect as *D. immaculata* possesses a stipe as well (Fig. 33B).

Two further structures of the *Huernia* gynostegium, which are only marginally mentioned by LEACH (1988), the double guide rail structure (cf. KUNZE 1982) and the hump-like outgrowth of the filament tube below the guide rail entrance (Fig. 18B), seem equally important for the generic definition. The latter character has to be viewed as one of the most important autapomorphies of *Huernia*. Both characters are of considerable importance for reproductive biology, because they effect a morphological incompatibility of foreign pollinia with the pollination apparatus of *Huernia* (compare JAHNKE 1989). The outgrowth in front of the guide rails practically excludes foreign pollinia and represents an almost perfect prezygotic isolation unequalled in stapeliads.

The guide rails, in addition, are characterized by being differentiated into an outer groove and an inner tube in *Huernia* (KUNZE 1982), while the guide rails in *Duvalia* are of a simple structure.

The similar structure of the pollinaria in both genera (Figs. 10 and 18C, D) is probably symplesiomorphic. It is significant that both in *Duvalia* and *Huernia*, but also in *Orbea*, in contrast to the other stapeliad genera, almost no change in the position of the pollinia after the extraction and the drying of the pollinaria can be observed.

These comparisons show that *Huernia* and *Duvalia* can be safely separated by a set of characters, despite many similar or even identical elaborations. At the same time, they are large, homogeneous species groups – *Huernia* is the largest genus in the *Stapelieae* s. str. and comprises more than 60 species. Despite the disjunct distribution of *Duvalia*, the present distribution of both genera from the southwestern Cape through East Africa to southwestern Arabia shows that the two genera have probably evolved in parallel from the same ancestors and should be regarded as sister groups.

The ecological potential of a taxon has as much part in the extension of its area as its phylogeny and the paleo-ecology of the region (WALTER 1973). The biogeographical differences expressed by *Duvalia* and its sister genus *Huernia* in southern Africa should be considered from this starting point. Most *Huernia* species occur in the summer rainfall area of the Transvaal and the Eastern Cape, even in Angola three species are found (LEACH 1985). In the winter rainfall area, *Huernia* (as well as *Orbea*) is rare. In southern Africa, *Huernia* has developed habitat preferences different from *Duvalia*. According to its habitat and climatic preferences, *Huernia* most likely has been less affected by the effects of the climatic oscillations described above, a likely reason for its higher species diversity. As a main migration pathway a north-south corridor in the summer rainfall area of the eastern subcontinent might have influenced speciation predominantly.

Relationship with *Orbea*. Micromorphology of the stem epidermis shows many similarities in basic epidermal pattern and sculpturing between *Duvalia* and *Orbea* in its present circumscription (ALBERS & al. 1989, ALBERS & MEVE, unpubl.). The frequent occurrence of corolline annuli, translator wings and relatively stiff caudicles of the pollinaria (see above), as well as similar shapes of stem [*D. vestita* – *O. variegata* (L.) HAW.; *D. eilensis* – *O. prognatha* (P.R.O. BALLY) L.C. LEACH] and the same plagiotropous growth form, suggest that *Duvalia* and *Orbea* are phylogenetically related. Biogeographical aspects support this hypothesis, since *Orbea* also displays a disjunct distribution, showing even more distinct geographical coincidence with *Duvalia* than with *Huernia*. The distribution centre of *Orbea* is located in the northeastern parts of South Africa (MEVE, unpubl.), and the distribution gap is restricted to the area between Tanzania and Somalia. *Orbea prognatha* has survived as a paleoendemic and is the sole species of the genus in Somalia. There was no pronounced northward migration in *Orbea* because the genus is absent on the Arabian Peninsula.

Taxonomy

Taxonomic history

THUNBERG (1794) had knowledge of seven stapeliads belonging to different genera, but the first *Duvalia* species were described by MASSON (1796) with *Stapelia caespitosa, S. elegans* and *S. reclinata*. Based on these three taxa HAWORTH (1812) established the genus *Duvalia*, adding SIMS' (1803) *S. radiata*, and described *D. compacta, D. glomerata, D. laevigata* and *D. tuberculata* as new. HAWORTH'S new genus was named in honour of H.A. DUVAL, a French botanist, who previously named *Haworthia* in thc *Liliaceae*.

The acceptance of *Duvalia* as a well-delimited genus took some time. DECAISNE (1844) placed it under *Stapelia* as a section. In his account, DECAISNE (1844) added 5 taxa, all of which became later synomyms of *D. caespitosa* and *D. elegans*. During his outstanding activities at Kew (1873–1934), N.E. BROWN started the first modern treatment of *Duvalia*. He clearly recognized that Haworth's genus formed a homogeneous and natural group. He combined HOOKER'S (1874) *Stapelia corderoyi* with *Duvalia*, and published *D. polita* (BROWN 1876) and *D. angustiloba* (BROWN 1883). Later, in his most important treatment of stapeliads in the "Flora Capensis", BROWN (1908) gave a more precise circumscription of *Duvalia*, recording the most important diagnostic characters: "corolla with annulus", "corona stipitate" and "translator winged". Here, he also described five new and distinctive species from the Cape Province.

BERGER (1910) adopted BROWN'S (1908) concept, as did WHITE & SLOANE (1937) with only slight alterations. BAYER (1977, 1984a) presented again a review of the taxa of southern Africa. In contrast to the treatments of BROWN and WHITE & SLOANE, he suggested only a considerably reduced number of species should be accepted for South Africa. COURT (1981) took over BAYER'S ideas.

LAVRANOS' extensive collecting activities in East Africa and Arabia brought five new taxa between 1969 and 1983 to light, and documented that *Duvalia* constitutes a significant geo-element of the Eritrean-Arabian province.

In a preliminary revision of the genus, MEVE & ALBERS (1990b) subdivided the genus into sect. *Duvalia* and sect. *Arabica* MEVE & F. ALBERS. Their conspectus is still valid, except for *Duvalia sulcata* s. l., which is now subdivided into subspecies. In addition, one new species, *D. gracilis*, is described in the present paper.

BOELE & al. (1987, 1990) listed 21 validly described species. In the present paper 17 species plus three infraspecific taxa are accepted.

Taxonomic treatment

Duvalia HAW., Syn. Pl. Succ. 44 (1812) et Suppl. Pl. Succ. 13 (1819) ("*Duvallia*").
= *Stapelia* L. sect. *Duvalia* (HAW.) DECNE., in DC., Prodr. **8**: 661 (1844).
Type: *Duvalia elegans* (MASSON) HAW. = *Stapelia elegans* MASSON.

Plants small, mat-forming stem succulents of plagiotropic growth; stems glabrous, 4- to 6-angled, 1–10 cm long, with (1) podaria spreading, leaf rudiments triangular, 1–4 mm long, not persistent, bearing stipular glands (sect. *Duvalia*) or (2) podaria with persistent conical teeth (leaf rudiments), stipules on plants absent (sect. *Arabica*), latex clear. Inflorescences cymoid-thyrsic, developing successively one to several flowers in a dichasial or bostrychoid manner; flowers arising from near base to middle of young stems, without or with short peduncle; pedicels variable in length, 2–40 mm long; sepals 1–6 mm long, green, acute, glabrous. Corolla rotate, 5-lobed, 10–50 mm diam., central part raised to form an annulus supporting the outer corona, corolla lobes ± plane to replicate into vertical plates. Corona gynostegial, composed of an outer ring of fused staminal and interstaminal parts and of inner free staminal parts [C(is) + Cs]; outer ring forming a pentagonal to rounded disc covering the annulus partially or totally; staminal corona crown-like, lobes bilobed, inner parts tongue-shaped, small, incumbent on the anthers, outer parts ovoid to subulate, spreading or raised, blunt or acute. Gynostegium stipitate, anther wings simple, with broad cartilaginous margins, basally spreading wing-like, inner surfaces covered with glandular trichomes; spacious nectar cavities hidden in the filament tube beneath each guide rail entrance, opening of nectar cavity circular to ellipsoid; anthers subquadrangular and horizontal, incumbent on the stylar head, stylar head rather discoid, slightly convex, whitish; pollinia flattened ovoid, 270–550 µm long, 200–350 µm wide, with ca. 150–250 pollen grains each, with strong, straight pellucid margin along the distal half of the outer side of each pollinium; caudicles 80–200 µm long, with saucer-shaped widenings attached to the pollinium, terminally to subterminally inserted; corpusculum ovoid to ellipsoid, with flattened lateral wings, each wing up to 0.5 mm wide. Follicles 50–140 mm long, 4–10 mm diam., glabrous, containing 40–120 seeds; seeds flattened, ovoid, winged, 3–8 mm long, 2–5 mm wide, brown, coma 10–18 mm long, white.

Key to the sections

1. Stems green or uniformly red under exposure to sun, leaf rudiments triangular and flattened, stipular glands present (indistinct or absent in *D. immaculata* and *D. vestita*) . . . . I. sect. *Duvalia*
– Stems greyish-green and usually blotched with red-brown, leaf rudiments conical, stipular glands absent . . . . II. sect. *Arabica*

I. *Duvalia* Haw. sect. *Duvalia*

Description. Stems varying from globose to finger-like, 4- to 5-angled (6-angled in *D. polita* and *D. corderoyi*), green, turning to uniform red under exposure to sun. Leaf rudiments triangular, short-lived and with stipular glands (stipules indistinct or absent in *D. immaculata* and *D. vestita*). Outer parts of the inner corona lobes obtuse, ± rounded.

Distribution. Angola, Botswana, Malawi, Mozambique, Namibia, South Africa, Swaziland, Zambia, Zimbabwe.

Key to the species

1. Annulus ± glabrous 2
– Annulus with indumentum 11
2. Corolla yellow, 10–15 mm diam.. . . . . .9. *D. parviflora*
– Corolla brown, red to purplish, > 15 mm diam 3
3. Annulus with a strong central depression, completely enclosing the coronal disc 4
– Annulus not strongly depressed centrally, more or less at the same level as the coronal disc 5
4. Annulus and corolla lobes uniformly chocolate-brown, annulus 8–11 mm diam., 2–3 mm high, corona straw-yellow . . . . 6. *D. immaculata*
– Annulus whitish with reddish spots, corolla lobes reddish brown, annulus 5–8 mm diam., 1.5–2 mm high, corona canary-yellow . . 7. *D. maculata*
5. Annulus 3.5–4.5 mm diam., corona white . . . . 1. *D. angustiloba*
– Annulus >5 mm diam., corona bright yellow, light brown to dark purplish 6
6. Corolla lobes incompletely replicate 7
– Corolla lobes ± completely replicate. 9
7. Annulus <2 mm high; margins of corolla lobes with thin, often bent trichomes, 1.5–2.5 mm long . . . . .8. *D. modesta*
Annulus >2 m high; margins of corolla lobes glabrous or with trichomes. 8
8. Corolla lobes bright purplish or dark red, margins with 2–3 mm long simple trichomes; corona bright yellow . . . . 10. *D. pillansii*
– Corolla dark brown or reddish brown, margins with simple or vibratile clavate hairs; corona brownish to reddish brown. . . . . 11. *D. polita*
9. Annulus <1 mm high; corona light brown or greenish-brown . . . . 5. *D. gracilis*
– Annulus >1mm high; corona brownish to red-brown (purple). 10
10. Annulus 2–4.5 mm high, corolla lobes 6.5–9 mm wide, the margins with trichomes at least 0.2 mm long . . . 2a. *D. caespitosa* var. *caespitosa*
– Annulus ca. 2 mm high, corolla lobes 4–6.5 mm wide, the margins glabrous or nearly so . . . . 2b. *D. caespitosa* var. *compacta*
11. Corolla with indumentum ± over entire adaxial surface 12
Corolla with indumentum limited to annulus and margins of corolla lobes 14
12. Corolla adaxially pubescent (sometimes missing on apices of corolla lobes), hairs whitish, at least 1 mm long, basal epidermis dull; corolla lobes ± completely replicate . . . . 12. *D. pubescens*
– Corolla pilose, hairs purplish, >1 mm long, basal epidermis brilliant, corolla lobes not or incompletely replicate. 13
13. Flowers 15–22 mm diam., corolla lobes spreading, corolla hairs blackish purple, leaf rudiments 1–1.5 mm long . . . . 4. *D. elegans*
– Flowers 20–30 cm diam., corolla lobes replicate at least half way down, corolla hairs purplish, leaf rudiments ≤ 1 mm long. . . . . . 13. *D. vestita*

14. Flowers 30–45 mm diam., corolla lobes ca. 12 mm wide, annulus covered with hairs ca. 3 mm long . . . . 3. *D. corderoyi*
– Flowers smaller, corolla lobes 6.5–9 mm wide, annulus nearly glabrous to densely covered with hairs at most 1 mm long . . . . 2a. *D. caespitosa* var. *caespitosa*

1. ***Duvalia angustiloba*** **N.E. Br.**, Gard. Chron. **20**: 230 (1883).

Type species: "From the Karoo, near or on the way to the Diamond Fields", Dickson s.n. sub Barkly 33 (holo-, K!).

Description. Stems 8–30 mm long, 8–20 mm wide, 4- to 5-angled, ± ovoid to cylindrical, dark or grey-green, glabrous, leaf rudiments ca. 2 mm long, acute, with stipular glands. Inflorescence of 1–20 prostrate flowers developing gradually, primarily in dichasia, later also in bostrychoid manner, from the base to near the middle of the stems, scentless or with slight odour of excrement; pedicels 20–40 mm long, ca. 1 mm diam.; sepals 2–3 mm long, basally ca. 1 mm wide; flowers 15–22 mm diam. Corolla 15–22 mm diam., chocolate-brown, rarely maculated brown/green, lobes totally replicated, 6–9 mm long, ca. 2 mm high at the base, with margins touching each other along their whole length, margins glabrous or with elongated papillae towards the base, papillae 50–100 μm long; annulus circular to pentagonal, 3.5–4.5 mm diam. 0.4–0.8 mm high, basally slightly constricted, with papilla, 30–60 μm long, rim slightly undulating and not or slightly overtopping the coronal disc. Corona white, sometimes maculated with red-brown, coronal disc a small pentagonal fringe, ca. 3 mm diam., lobes of staminal corona 1.2–1.5 mm high, ca. 1.5 mm long. Pollinia ca. 270 μm long, 180 μm wide, orange-yellow, pellucid margin ca. 110 μm long; caudicels 80–100 μm long, translator wings bright orange, hyaline, ca. 220 μm long; corpusculum ca. 100 μm long. Follicles less than 10 cm long; seeds 4–5 mm long, 2.5–3 mm wide, wing ca. 500 μm wide, coma 10–13 mm long. Chromosome number: $2n = 22$ (vouchers: see * in "Specimens examined") (Fig. 19).

Distribution. Western and Eastern Cape, Great Karoo; 32°10′–33°10′S and 22°50′–24°20′E (Fig. 20).

Habitat. Typically found in the Acocks Veld Type 30, Central Lower Karoo, 800–1000 m.

This species is characterized by distinctive spider-like corolla and the (mostly) pure white corona. Apart from that, it is one of the few species with a clear eco-geographical preference expressed by its association to the veld type "Central Lower Karoo". Here, *D. angustiloba* is endemic to the flat, often pan-like habitats of the Great Karoo between Beaufort West and Aberdeen (Fig. 20). This area around 800–1000 m altitude mostly provides monotonous semi-arid sites on sand with small boulders, mostly covered with *Pentzia*, *Chrysocoma*, *Lycium*, or *Ruschia*.

Variability within *D. angustiloba* is low and restricted to the length of corolla papillae and corolla coloration.

Duvalia gracilis obviously is the nearest relative of *D. angustiloba*, but differs in corolla sizes, coloration and indumentum.

Specimens examined

South Africa:

3222 AD: Stolshoek (Karoo Natl. Park), Bruyns 3417 (BOL).

3222 DC: Trakaskuil, Bruyns 3678 (BOL).

3222 DD: between Prince Albert and Rietbron, "Prutkraal", Meve 316* (MSUN); Bruyns 3148 (NBG).

3223 AA: Nelspoort, sub KG 771/60 (NBG!).

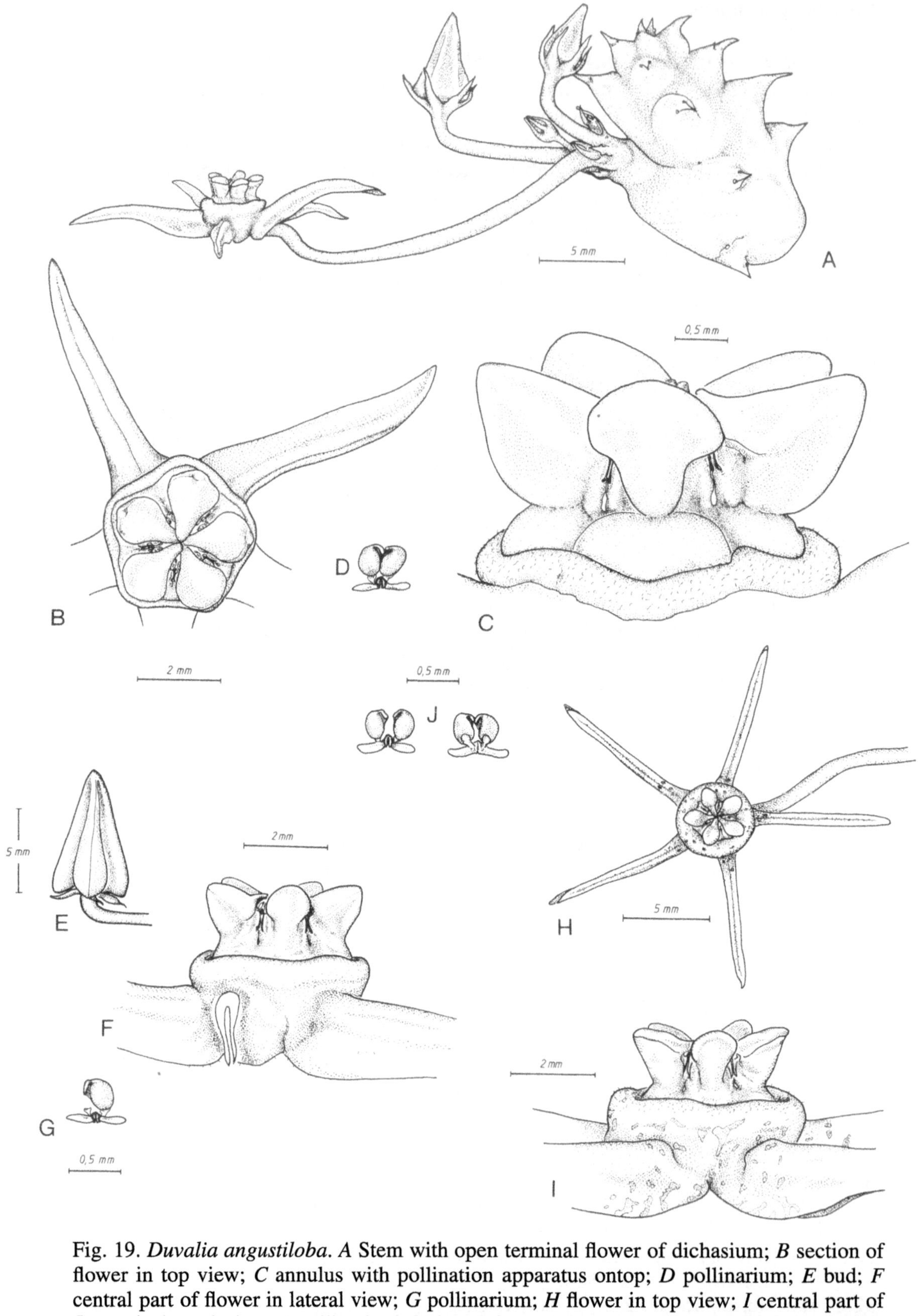

Fig. 19. *Duvalia angustiloba*. *A* Stem with open terminal flower of dichasium; *B* section of flower in top view; *C* annulus with pollination apparatus ontop; *D* pollinarium; *E* bud; *F* central part of flower in lateral view; *G* pollinarium; *H* flower in top view; *I* central part of flower in lateral view; *J* pollinaria, left dorsal face, right ventral face. (*A* Meve 320; *B-D* Meve 316; *E–J* Meve 322)

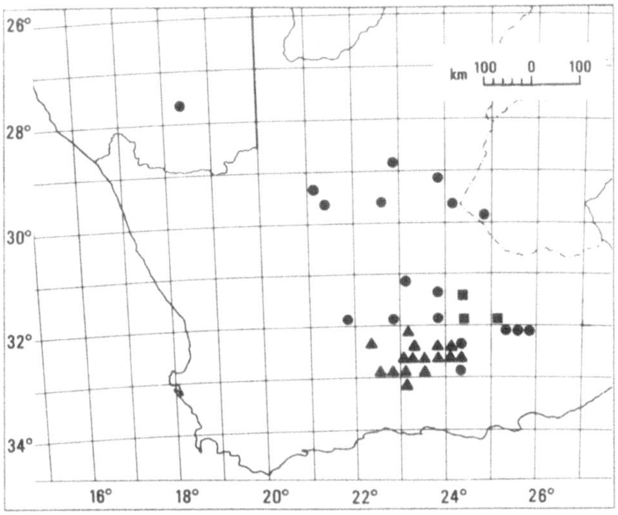

Fig. 20. Distribution of *Duvalia angustiloba* (▲), *D. gracilis* (■) and *D. maculata* (●)

3223 BD: 28 km W Aberdeen, MEVE 324* (MSUN).
3223 CA: 35 m Beaufort West - Aberdeen, BAYER 2097 (NBG!); Rooidam, BRUYNS 2832 (BOL)
3223 CB: 82 km E Beaufort West, MEVE 319* (NBG), MEVE 320* (MSUN).
3223 CC: Rietbron, BRUYNS 3135 (NBG).
3223 DA: 112 km E Beaufort West, BRUYNS 1755 (NBG!); 91 km E Beaufort West, MEVE 321*, 322*, 323* (MSUN).
3223 DB: 39 km from Aberdeen on road to Willowmore, PERRY 658 (NBG!).
3224 AC: 12 km E Aberdeen, BRUYNS 2957 (BOL).
3224 CA: Wallacedale, BRUYNS 4244 (BOL).
3224 CB: Marais, BRUYNS 2965 (BOL).
3323 AA: 15 km S Rietbron, BRUYNS 3132 (NBG).

2. ***Duvalia caespitosa*** **(Masson) Haw.,** *Syn. Pl. Succ. 45 (1812). Basionym: Stapelia caespitosa* Masson, Stap. Nov., 20, t. 29 (1796).

Type: Icon. Masson, Stap. Nov., t. 29 [lecto-, designated by N.E. Brown (1908)].

Key to the varieties of *D. caespitosa*

1. Annulus 2–4.5 mm high, papillate or hairy, corolla lobes 6.5–9 mm wide, margins with at least a few simple hairs, with or without vibratile clavate hairs . . . . 2a. var. *caespitosa*
– Annulus 1.2–2.5 mm high, ± papillate, corolla lobes 4–6 (6.5) mm wide, margins glabrous (rarely with some short, simple hairs) 2b. var. *compacta*

2a. ***Duvalia caespitosa* var. *caespitosa***

= *Stapelia caespitosa* var. *hirtella* (Jacq.) Loud., Encycl. Pl., 202 (1841). Type: Icon. Jacq., Stap., t. 10.

= *Duvalia radiata* (Sims) Haw., Syn Pl. Succ. 45 (1812). Basionym: *Stapelia radiata* Sims, Bot. Mag., t. 619 (1803) [nec Jacq. 1806–1819]. Type: Icon. Sims, Bot. Mag., t. 619 [lecto-, designated by N.E. Brown (1908)].

= *Duvalia radiata* var. *minor* (N.E. Br.) A.C. White & B. Sloane, Stap. **2**: 781 (1937). Basionym: *Duvalia hirtella* var. *minor* N.E. Br. in Thiselton-Dyer, Fl. Cap. **4**(1): 1031 (1908). Type: Worcester Distr., Pillans 628 (holo-, K, not found).

= *Duvalia radiata* var. *obscura* (N.E. Br.) A.C. White & B. Sloane, Stap. **2**: 778 (1937). Basionym: *Duvalia hirtella* var. *obscura* N.E. Br. in Thiselton-Dyer, Fl. Cap. **4**(1): 1031 (1908). Type: Karoo near Matjiesfontein, 13 April 1903, Pillans 13 (lecto-, K!, designated here).

= *Duvalia radiata* var. *hirtella* (Jacq.) A.C. White & B. Sloane, Stap. **2**: 779. Basionym: *Stapelia hirtella* Jacq., Stap., t. 10 (1806–1819). Type: Icon. Jacq., Stap., t. 10 [lecto-, designated by Loudon (1841)].

= *Duvalia hirtella* (Jacq.) Sweet, Hort. Brit., 1st edn. 1, 276 (1827). Basionym: *Stapelia hirtella* Jacq., Stap., t. 10 (1806–1819). Type: Icon. Jacq., Stap. t. 10 [lecto-, designated by Loudon (1841)].

= *Duvalia hirtella* var. *minor* N.E. Br. in Thiselton-Dyer, Fl. Cap. **4**(1): 1031 (1908). Type: Worcester Distr., Pillans 628 (holo-, K, not found).

= *Duvalia reclinata* (Masson) Haw., Syn. Pl. Succ., 44 (1812). Basionym: *Stapelia reclinata* Masson, Stap. Nov., 19, t. 28 (1796) [non *Stapelia reclinata glauca* Haw., in schedis]. Type: Icon. Masson, Stap. Nov., t. 28 (lecto-, designated here). - *D. pubescens* N.E. Br. p.p., Court, Succ. Fl. S. Afr., 198 (1981); M.B. Bayer, Veld Fl. **70**: 13 (1984).

= *Duvalia reclinata* var. *bifida* N.E. Br. in Thiselton-Dyer, Fl. Cap. **4**(1): 1030 (1908). Type: Eastern Cape Province, Glen Avon Estate (2 miles east of) near Somerset East, 5 May 1906, Pillans 27 (holo-, K!; iso-, BOL).

= *Duvalia reclinata* var. *angulata* N.E. Br. in Thiselton-Dyer, Fl. Cap. **4**(1): 1030 (1908). Type: Pillans 615 (holo-, K, not found).

= *Duvalia replicata* (Jacq.) Sweet, Hort. Brit., 1st edn, 276 (1827). Basionym: *Stapelia replicata* Jacq., Stap., t. 15 (1806–1819) Type: Icon. Jacq., Stap., t. 15 (lecto-, designated here).

= *Duvalia emiliana* A.C. WHITE, in A.C. WHITE & B. SLOANE, Stap., 1st edn, 126 (1933). Type: Icon. WHITE & SLOANE, Stap. **2**: 763, (lecto-, fig. 760, photo, designated here); origin of type plant (WHITE & SLOANE 143) not known.
= *Duvalia propinqua* A. BERGER, Monatsschr. Kakteen, 24 (1904). Type: Cultivar La Mortola, BERGER s.n. (holo-, K!).
= *D. marlothii* N.E. BR., *nomen nudum, in schedis*. Type: Hort. Kew., 29 July, 1921 (holo-, K!). BROWN: "Plant received from Dr. MARLOTH without locality".
– *D. maculata auct. non* N.E. BR.: PHILLIPS, Fl. Pl. S. Afr. **12**, t. 474 (1932).

Description. Stems 15–100 (–130) mm long, 10–22 mm diam., 4- to 5-angled, ovoid to cylindrical, dark green, glabrous, leaf rudiments ca. 2 mm long, stipular glands relatively large. Inflorescence bostrychoid, 1 to several flowers developing successively, flowers with weak to strong excrement odour. Pedicels 10–25 mm long, 1.5–2.5 mm diam.; sepals 3–6 mm long, 1–2.5 mm wide at the base. Corolla 20–35 mm diam., bright reddish-brown, chocolate, brown-purple to dark purple, lobes 9–15 mm long, 6–9 mm wide, almost completely replicate, with trichomes ca. 200–400 µm long and/or vibratile, clavate hairs, (200–) 600–2600 µm long along the margins either basally or for most of their length; annulus circular to pentagonal, sometimes basally constricted, rim convex or flattened, 7–11 mm in diam., 2–5 mm high, centrally sometimes maculate with white, yellow or brown, variably covered with papillae and/or trichomes, 40–1000 µm long. Coronal disc circular to pentagonal, 5–7 mm in diam., bright reddish-brown, bright orange-brown or purplish to purple-brown, dorsal lobules of staminal corona 1.5–2.2 mm long, canary-yellow, yellow, brown-yellow, brown, reddish-brown, orange-brown or brownish-purple. Pollinia 350–500 µm long, 250–300 µm wide, dark orange-brown or yellow, rarely yellowish-green, pellucid margin 180–230 µm long; caudicles 100–150 µm long, orange, translator wings whitish to bright orange-brown, 350–450 µm long; corpusculum ca. 200 µm long. Follicles 6–17 cm long; seeds 4–5.5 mm long, ca. 3 mm diam., seed rim 400–800 µm wide, coma 12–18 mm long. Chromosome number: 2n = 44 (vouchers: see * "Specimens examined") (Figs. 21, 22).

Distribution. Western and Northern Cape (Great and Little Karoo) Eastern Cape, Free State; 29°20′–34°55′ S and 18°40′–27°30′ E (Fig. 23).

Habitat. Widely distributed in the transition zone of summer and winter rainfall areas and in the summer rainfall area of mainly the Western and Eastern Cape and rarely in the Free State. Exhibiting a wide ecological amplitude, but mainly found in Acocks Veld Types 25 and 26, Succulent Mountain Scrub and Karroid Broken Veld; alt: 100–1400 m a.s.l.

The history of this species starts with MASSON's (1796) description of *Stapelia reclinata* as distinct from *S. caespitosa*. MASSON differentiated between both taxa because of the occurrence of clavate hairs. This character, however, is not of significance in var. *caespitosa*, because it is variable even within single populations.

The large number of specific names puplished for this species during the last 200 years is not surprising considering its enormous variability in flower morphology. *Duvalia caespitosa* possesses almost the largest number of synonyms within the stapeliads and is surpassed only by *Orbea variegata* (L.) HAW. (cf. LEACH 1978).

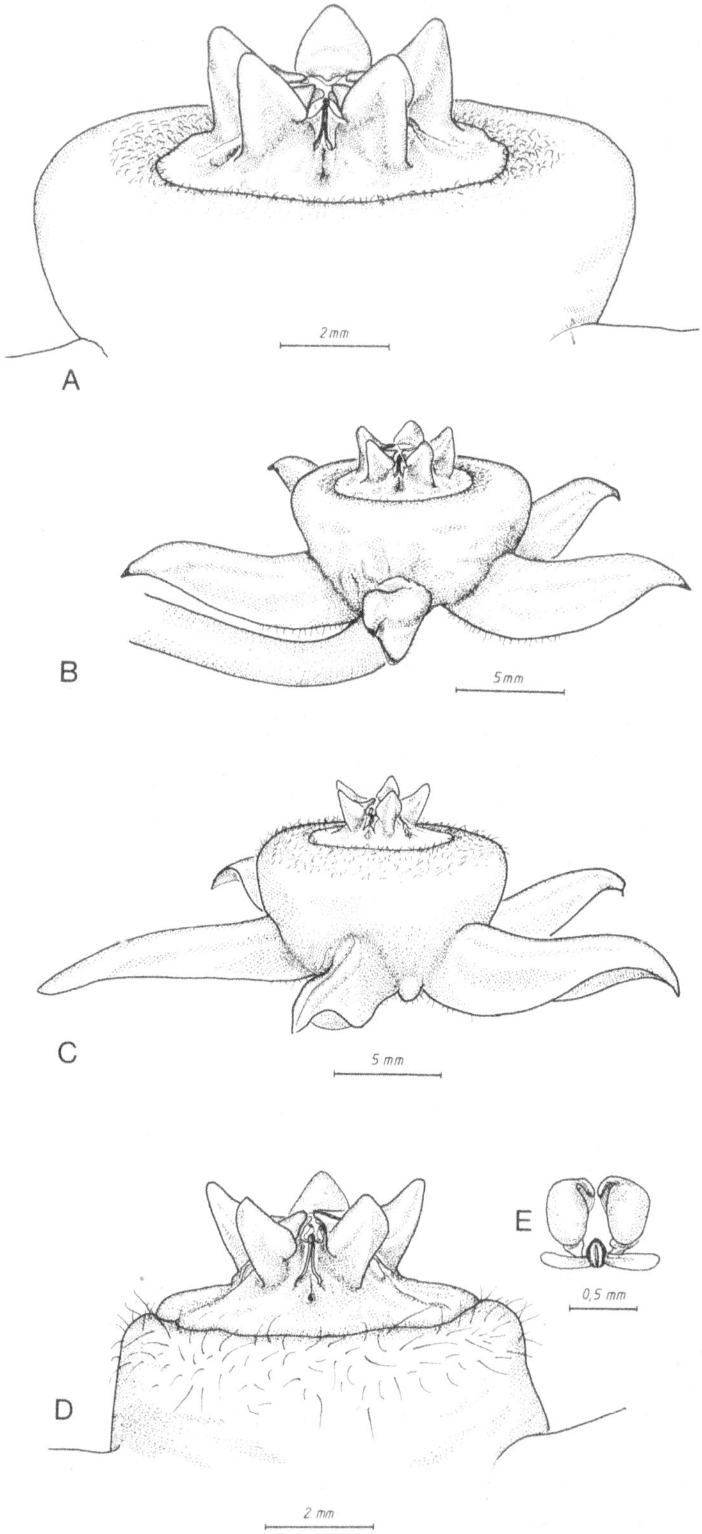

Fig. 21. *Duvalia caespitosa* var. *caespitosa*. *A–D* Flowers in lateral view; *E* pollinarium. (*A*, *B* ALBERS & al. sub K 1389; *C* MEVE 390; *D*, *E* KUSCH & MEVE sub K 1524)

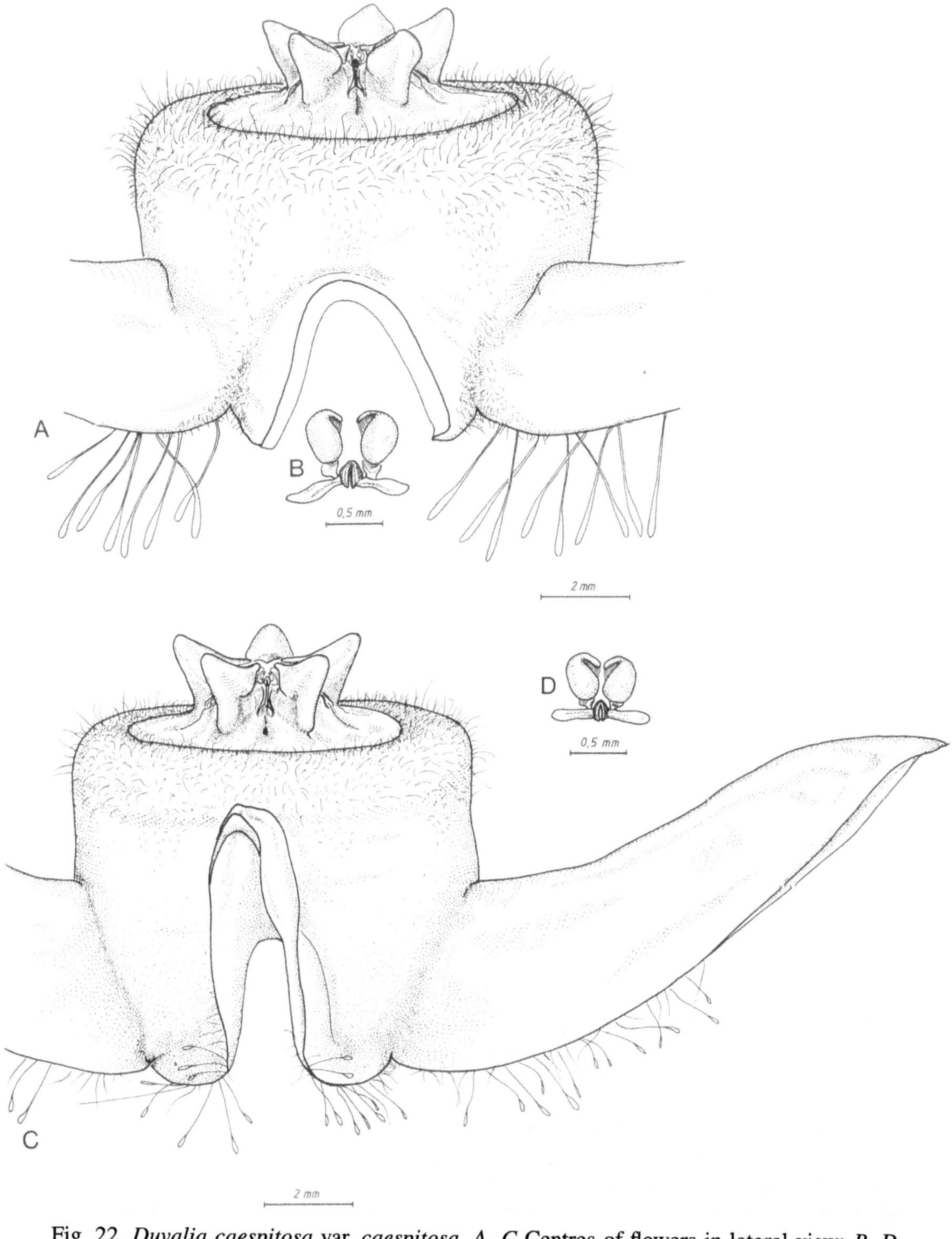

Fig. 22. *Duvalia caespitosa* var. *caespitosa*. *A*, *C* Centres of flowers in lateral view; *B*, *D* pollinaria (*A*, *B* ALBERS s.n. sub K 1117; *C*, *D* MEVE 404)

However, *D. elegans* has been confused with *D. caespitosa* in the past since JACQUIN described true *D. elegans* as *D. radiata* and published *D. radiata* Sims as *D. replicata*. For more remarks see under *D. caespitosa* var. *compacta* and *D. elegans*.

Specimens examined

South Africa:

2615 DA: Namibia, Halenberg, STAYNER sub KG 674/61 (NBG!).
2925 CB: Fauresmith, Bergplaas, VERDOORN 1350 (PRE!).
2925 CC: SW Fauresmith, Farm Varkfontein, GERBAULET & STRUCK 23745.
3023 DA: Twyfelhoek, BRUYNS 3022 (BOL).
3027 CC: Cape, Wodehouse, Clanville, Dordrecht, BAYLISS 2683 (NBG!).
3120 DC: 13 km W Luzernvlei, BRUYNS 4008 (BOL).
3125 CC: 66 km W Cradock, MEVE 461* (MSUN).
3220 AB: $63\frac{1}{2}$ km from Sutherland to Williston, on old Calvinia Rd, BAYER 2170 (NBG!); Cape, 13 mi SW Middlepos Wolwedans, LEACH & BAYER 16065 (NBG!).
3220 AD: Sutherland, Voelfontein Farm, HALL 219 (NBG!); Cape, Sutherland, Ouberg-Pass, LAVRANOS & PEHLEMANN 18841* (MSUN).
3220 BA: between Sutherland and Williston, BAYER 2143 (NBG!).
3220 BC: 10.5 km N Sutherland, Kruis River, BAYER 643 (NBG!);
3220 CB: 2 km from the Bowadrif turnoff on the Karoopoort- Sutherland Rd, BAYER 759 (NBG!).
3220 DB: Merweville Rd, Bloukop on farm Harde, WILLIAMS 130 (NBG!)
3220 DD: 52 km Moordenaars K. from Laingsburg, BAYER 2130 (NBG!).
3221 CC: Klipfontein, BRUYNS 3111 (NBG).
3221 CB: 24 km W Merweville, BAYER 2469b (NBG!).
3222 CD: Klein Waterfal, BRUYNS 3800 (BOL).
3224 AB: NW Graaf Reinet, St. Olives, Ouberg, BAYER 2352b (NBG!); NW Graaff Reinet, Ouberg Pass, MEVE 326* (MSUN); NW Graaff Reinet, "St.Olives" (W slope of Ouberg), MEVE 339*, 340 (MSUN).
3224 BA: N Graaff Reinet, BAYER 2025 (NBG!); 23 km NE Graaff Reinet, BRUYNS 1762 (NBG!).
3224 BC: Graaff Reinet Commonage, BAYER 2369 (NBG!).
3224 CB: Aberdeen Rd., BRUYNS 4249 (BOL).
3224 CD: between Klipplaat and Aberdeen Road (32°54′ S 24°16′ E), ALBERS & al. 10618 sub K 1122* (MSUN); 10 km N Klipplaat, BRUYNS 4251 (BOL).
3224 DB: Pearston, Cranemere, BRUYNS 1796b (NBG!); W Pearston, "Cranemere", MEVE 440*, 441*, 442* (MSUN).
3225 AC: ca. 20 km NE Pearston (top of Buffelshoek Pass), MEVE 454*, 455* (MSUN).
3225 BA: Cradock, Greenacres, BRUYNS 1574 (NBG!).
3225 CA: Pearston Skietbaan, BRUYNS 1588, 1588b (NBG!); 4 km S Pearston, MEVE 433* (MSUN); 9 km N Pearston, MEVE 453* (MSUN).
3225 CD: 18 km W Milton, BAYER 739 (NBG!).
3225 DA: Somerset East, MAC OWAN 2232 (K!)..
3225 DD: 2 km S Middleton, BAYER 649 (NBG!).
3226 CA: SW Bedford, "Sheldon Station", PRINGLE s.n. sub PLOWES 2989 (SRGH).
3319 BA: 4 km N Karoopoort, LAVRANOS & PEHLEMANN 17468* (MSUN).
3319 BB: 1 km N Karoopoort, ALBERS & al. K 1414* (MSUN); Ceres Karoo, Karoopoort, WISURA 2112 (NBG!).

3319 BC: De Doorns, BAYER s.n. sub KG 82/73 (NBG!).
3319 BD: De Doorns, Tunnel Stasie, BAYER s.n. (NBG!).
3319 CB: 3.5 km SE Aan de Doorns, MEVE 388* (MSUN); E Worcester, E Meiring Park, FORRESTER 111 (NBG!); Worcester, S Veldreservaat, OLIVER 218 (STEU).
3319 CD: 15 mi S Worcester, Lemoenpoort, BAYER 645 (NBG).
3319 DA: 8.5 km SE Worcester, Sandberg Hills, BRANDHAM 3071* (K); 13 km SE Worcester, BRUYNS & BAYER 3/75 (NBG!); 13 mi SE Worcester, Mowers, BAYER 651 (NBG!); 18 km SE Worcester, BAYER 1412 (NBG!); Keeromdam, Koo, near farm Witwater, HIEMSTRA 295 (NBG!); 5 km SE Aan de Doorns, MEVE 390*, 391* (MSUN).
3319 DB: Langvlei, on the way to Buitenstekloof, ALBERS & al. K 1378* (MSUN).
3319 DC: Robertson, Farm Dublin, STAYNER sub KG 399/61 (NBG!).
3319 DD: between Langvlei and Voorspoed, ALBERS & al. K 1382* (MSUN); Langvlei, ca. 10 km NW Robertson, MEVE 377*; 378* (MSUN).
3319 DD: McGregor, BAYER 765 (NBG!); ca. 5 km SE McGregor, MEVE 385*, 386* (MSUN).
3320 AB: W Matjiesfontein, Jagerskraal, ALBERS & al. K 1389*; NW Matjiesfontein, S Palmietfontein, ALBERS & al. K 1391* (MSUN); S Palmietfontein "Jagerskraal", MEVE 295* (MSUN); Jagerskraal, MEVE & KUSCH K 1522* (MSUN); Tweedside, s. coll. s.n. NBG 642/51 (NBG!); Pieter Meintjes, BAYER 652 (NBG!); N Pieter Meintjies Station, BAYER 2101 (NBG!).
3320 AC: NE Avondrust, Touwsriver, BAYER & STAYNER 678 (NBG!).
3320 AD: ca. 37 km SE Touwsrivier (rd. to Bloutoring), MEVE 291, 292*, 293* (MSUN).
3320 BA: 2 km NW Matjiesfontein, ALBERS & al. K 1421*, K 1424* (MSUN); 500 m E Matjiesfontein, GERBAULET & STRUCK 24291* (MSUN); 25 km W Laingsburg, Matjiesfontein, LAVRANOS 21914 (MSUN)*; LAVRANOS & PEHLEMANN 20643 (MSUN)*; Matjiesfontein, LITTLEWOOD s.n. (NBG!); near Matjiesfontein, PILLANS 13 (K!); Matjiesfontein, PILLANS 81 (K!); Matjiesfontein, STAYNER sub NBG 481A/59 (NBG!).
3320 BB: Base of Witteberg, WALGATE 253 (NBG!).
3320 BD: ca. 28 km S Laingsburg, Rooinek Pass, ALBERS 2482* (MSUN); 20 km SW Laingsburg on Agter Witteberg Rd, BAYER sub KG 73/72 (NBG!); 14 km SW Rooinek-Pass, BAYER 637 (NBG!); 32 km from Ladismith to Laingsburg, BAYER 669 (NBG!); 4 km SE Joubertskop, BAYER 2705 (NBG!); S Laingsburg, Keurfontein, BAYER 4670 (NBG!); E Keurfontein, BAYER 5033 (NBG!); S Laingsburg, Keurfontein, BAYER s.n. sub KG 383/77 (NBG!); between Laingsburg and Ladismith, MEVE & KUSCH K 1552* (MSUN); Laingsburg, Rooinekpass, LAVRANOS & PEHLEMANN 17549; ca. 10 km SW Laingsburg (Witteberge); LAVRANOS & PEHLEMANN 20897* (MSUN); 30 km SW Laingsburg (to Ladismith), MEVE 284* (MSUN); 20 km S Laingsburg, MEVE 370, 371* (MSUN); ca. 18 km SW Laingsburg (Witteberge), MEVE 283* (MSUN).
3320 CA: Bonnievale, Goudmyn, BAYER 679 (NBG!).
3320 CC: NE Montagu, Rietvlei no.2, BAYER 744 (NBG!); 8 mi E of Montagu, HALL 877 (NBG!); Montagu, Rietvlei no.1, JOUBERT s.n. sub KG 278/73 (NBG!); Bonnievale, Boesmans River, ROUSSOUW 439 (NBG!); Montagu Bath, WISURA 476 (NBG!).
3320 DB: Plathuis (between Barrydale and Ladismith), BAYER 641 (NBG!).
3320 DC: 2 km W Barrydale, BAYER 460 (NBG!); Barrydale, Warmwaterberg, BATTEN 479 (SRGH); 3 km S Warmwaterberg (between Barrydale and Ladismith), BAYER 640 (NBG!).

3320 DD: between Muiskraal and Barrydale (near Kleindoornrivier), ALBERS & MEVE 129* (MSUN).
3321 AD: Ladismith, GROSSKOPF s.n. (STEU); Amalienstein (Little Karoo), s. coll. s.n. (STEU 6930, STEU).
3321 BC: Calitzdorp Dam/Besemkop, MEVE & KUSCH K 1556* (MSUN); Calitzdorp, Huisrivierpass, LAVRANOS & PEHLEMANN 20629* (MSUN); Calitzdorp Dam, s. coll. s.n. NBG 1278/49 (NBG!).
3321 BD: Groenfontein near Calitzdorp, LAVRANOS & PEHLEMANN 19785 (MSUN)*; 2 km W Calitzdorp, LAVRANOS & PEHLEMANN 20626*; Calitzdorp Dam, s. coll. (NBG 1274/49, NBG!).
3321 CA: ca. 25 mi S Ladismith, BRUYNS 1643b (NBG!); S Ladismith, Ockertskraal, BRUYNS s.n. (NBG!).
3321 CB: 4 km SW Ladismith, ALBERS 3022 et 3023* (MSUN).
3321 CD: Vanwyksdorp area, STAYNER s.n. sub KG 97/70 (NBG!).
3321 DA: 8 mi S Calitzdorp, BAYER 434 (NBG!); Van Wyksdorp, BAYER 3748 (NBG!); 8 mi S Calitzdorp, BAYER s.n. sub KG 130/72 (NBG!).
3321 DB: Middlepad to Oudtshoorn and Calitzdorp, STAYNER sub 864/60 (NBG!); between Middlepad and Oudtshoorn /Calitzdorp Rd, sub s. coll. s.n. KG 927/ 60 (NBG!).
3321 DC: 10 mi SE Vanwyksdorp, BAYER 667 (NBG!).
3322 AA: near Prince Albert, MARLOTH 4585 (K!); Prince Albert, SLABBERT s.n. (STEU 5942, STEU).
3322 AB: 29 km E Prince Albert, MEVE 308*; 16 km E Prince Albert, MEVE 309*; Tierberg, BRUYNS 2810 (BOL).
3322 AC: 5 km S Cango Caves, BAYER s.n. sub KG 383/75 (NBG!); Schoemans Kloof, WISURA 1794 (NBG!)
3322 BC: Klaarstroom, LAVRANOS & PEHLEMANN 20951* (MSUN).
3322 CA: Oudtshoorn, TAYLOR 712/29 (K!).
3322 CB: ca. 20 km SE Oudtshoorn, MEVE 403*, ca. 19 km SE Oudtshoorn, MEVE 404*, ca. 18.5 km SE Oudtshoorn, MEVE 411*, 18 km SE Oudtshoorn, MEVE 412* (MSUN); 1.5 km S Dysseldorp, MEVE 418* (MSUN).
3322 CC: Oudtshoorn, 2–6 km N Moeras River, BAYER 541 (NBG!); 19 km SW Oudtshoorn, BAYER s.n. sub KG 147/71 (NBG!).
3322 DA: 13 km E De Rust, BAYER 646 (NBG!); 12 km E De Rust, MEVE 419, 420* (MSUN); 30 km E Oudtshoorn, PERRY 1465 (NBG!).
3322 DB: E De Rust, Buffelsklip, BAYER 648 (NBG!); Buffelsklip, between Uniondale and De Rust, BAYER s.n. sub KG 214/74 (NBG!); 5 km W Buffelsklip, LAVRANOS & PEHLEMANN 17546* (MSUN); Uniondale, Buffelsklip, BAYER 664 (NBG!).
3323 AD: 20 km W Willowmore, BAYER sub KG 207/74 (NBG!); Willowmore, BRAUNS s.n. (STEU).
3323 BD: Vryfontein, ROUSSOUW 513 (NBG!).
3323 CA: Uniondale, FOURCADE 4638 (BOL); 16 km Uniondale on De Rust road, ex HAGNER s.n. sub PLOWES 3779 (SRGH).
3323 DA: 44 km Willowmore to Baviaanskloof, ROUSSOUW 140 (NBG!).
3324 AA: 10 km from Klipplaat to Willowmore, BAYER 650 (NBG!).
3324 AB: 2.5 km S Mount Stewart, MEVE 427* (MSUN); 15 km S Klipplaat, ROUSSOUW 203 (NBG!).
3324 BB: W Waterford, BAYER 4967 (NBG!).
3324 BD: E Steytlerville, Springbokvlakte, BRUYNS 1827 (NBG!).

3324 CA: Dam se Drif, BRUYNS 1847 (NBG!); on koppie W Latti homestad, ROUSSOUW 477 (NBG!); Sandvlakte, STAYNER s.n. sub KG 193/71 (NBG!); 20 mi S Studtis, Coega, BRUYNS 2193 (NBG!).
3324 CB: Steytlerville, Kouga Dam, STAYNER s.n. sub KG 370/62 (NBG!).
3324 DD: 2 km E Hankey, BAYER 1930 (NBG!); 1 km E Hankey on old road, BAYER s.n. sub KG 178/73 (NBG!); 1 m NE Hankey, BRANDHAM & CUTLER 121/1393 (K!); 2 km from Hankey on road to Uitenhage, PERRY 657 (NBG!).
3325 AC: Sapkamma, STAYNER sub KG 204/60 (NBG!).
3325 AD: Kirkwood, Sunday River, PLOWES 5187 (SRGH); 13 km SW Kirkwood, Steenbokvlakte, PLOWES 5235 (SRGH).
3325 BB: Kommadagga, Verdun, BRUYNS 1568 (NBG!).
3325 CB: between Uitenhage and Steytlerville (33°35'S 25°20'E), ALBERS & al. 10588 sub K 1117* (MSUN).
3325 DA: Bluewaterbay, Addo Bush, BATTEN 256 (SRGH); 5 mi S Addo, Soutkloof, BRUYNS 1624 (NBG!); Soutkloof, BRUYNS 4261 (BOL).
3325 DC: 2 km E Coega, BRUYNS 2169 (NBG!).
3326 BB: Commitees Drift (East Bank, Fish River), BAYER 414 (NBG!); 35 ml W Grahamstown, Keiskamma, BAYER s.n. sub KG 131/73 (NBG!).
3326 BC: Grahamstown, MAC OWAN 2232 (SAM!); Kowie River, STAYNER s.n. sub KG 246/69 (NBG!).
3327 AA: Keiskamma River, Peddie, s. coll. (NBG 463/50, NBG!).

Localities unknown or imprecise: SOUTH AFRICA, Cape, Karoo, BARKLY 51 et 53 (K!); E Cape, between Mosselbai -Pt.Elizabeth, GERWITZ s.n. (STEU 5903, STEU); SW Cape, Versene Bok (ca. 15 km W of Kraal to Laingsburg) LAVRANOS & PEHLEMANN sub K 1716* (MSUN); LAVRANOS & PEHLEMANN 18851a*, 19716* (MSUN); PILLANS 73 (K!). 3219, Cedarberg, s. coll. s.n. sub NBG 13/43 (NBG!). 15 km W Kraal, LAVRANOS & PEHLEMANN 19703* (MSUN).

2b. ***Duvalia caespitosa* var. *compacta* (HAW.) MEVE, comb. et stat. nov.**
Basionym: *Duvalia compacta* HAW., Syn. Pl. Succ., 46 (1812).

Type: The specimen in the HAWORTH Herbarium (OXF) labelled as "*Duvalia compacta*" was clearly seen by HAWORTH, and obviously represents the holotype.

= *Stapelia compacta* SCHULT., in H. ROEMER & SCHULT., Syst. Veg. **6**: 46 (1820). Type as for *D. compacta*.

= *Duvalia mastodes* (JACQ.) SWEET, Hort. Brit., 1st edn, 276 (1827). Basionym: *Stapelia mastodes* JACQ., Stap., t. 13 (1806–1819); LINK, Enum. Pl. Hort. Berol. **1**: 258 (1821) (as *S. mustodes* amongst *S. punctata*); ST.-LEG., Ann. Soc. Bot. Lyon **7**: 135 (1880) (as *S. mastodis*). Type: Icon. JACQ., Stap., t. 13 (lecto-, designated here).

Description. Like the typical variety but differing in a considerable number of sometimes cryptic features:

Stems 15–70 (–90) mm long, 10–25 mm diam., green or dark green, leaf rudiments 1.5–4 mm long. Pedicels 8–28 mm long, 1–2 mm diam.; sepals 2–4 mm long, 1–1.5 mm wide at the base. Corolla 16–24 (–28) mm diam., red to chocolate-brown, lobes 8–13 mm long, 4–6 (–6.5) mm wide, margins glabrous or basally with trichomes, 50–100 (–600) µm long; annulus 5.5–8 mm diam., 1.2–2.5 mm high, upper part of the rim variably covered with hair papillae, 30–100 µm long. Coronal disc (3.5–) 4.5–5(–6) mm diam., dorsal lobules of staminal corona 1.4–

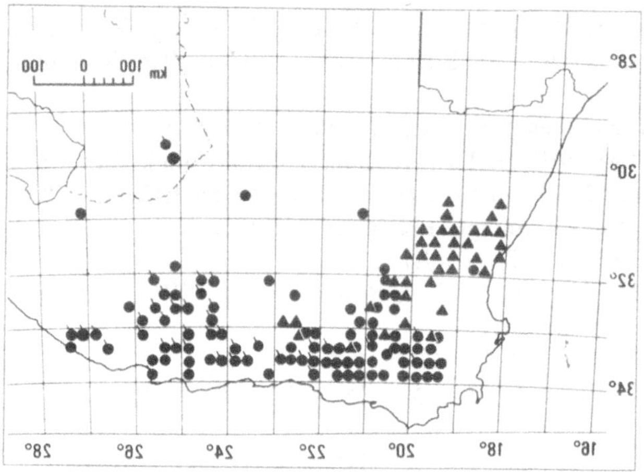

Fig. 23. Distribution of *Duvalia caespitosa* var. *caespitosa* without vibratile clavate hairs (dots) and with vibratile clavate hairs along corolla margin (●́) and of *D. caespitosa* var. *compacta* (▲)

2.2 mm long, light brown to brownish-purple. Pollinia orange-brown, 350–450 μm long, 220–300 μm diam., pellucid margin 140–200 μm long; caudicles 100–150 μm long, translator wings, 350–400 μm long; corpusculum ca. 180 μm long. Follicles 7–12 cm long; seeds 4.5–6 mm long, ca. 3 mm diam., wing ca. 500 μm wide, coma 12–16 mm long. Chromosome number: $2n = 44$ (vouchers: see * "Specimens examined"). (Figs. 24, 25).

Distribution. Western Cape, Great Karoo; 30°40′–33°20′S and 18°05′–22°55′E (Fig. 23).

Habitat. Rather exclusively limited to the winter rainfall area of the south-western Cape. Found mainly in the Succulent Karoo (Acocks Veld Typ 31), Western Mountain Karoo (VT 28) and the Namaqualand Broken Veld (VT 33), alt.: 50–1050 m a. s. l.

Haworth (*D. compacta*, 1812) and Jacquin (*Stapelia mastodes*, 1806–19) obviously described the species independently on the same plant material. The exact publication date for each of the originally unnumbered plates of Jacquin's "Stapeliarum in hortis vindobonensibus cultarum descriptiones figuris coloratis illustratae" is impossible to trace (cf. White & Sloane 1937, Rowley 1984). Even though the enumeration used for Kew's copy is widely accepted, the question of when exactly between 1806 and 1819 the single plate of *S. mastodes* was published is still not answered. Priority is given to *D. compacta*, because this epithet is in use and widely accepted because of White & Sloanes's explicit description and illustration.

The two varieties represent two ecotypes, even though individual "runaways" occur. *D. caespitosa* var. *caespitosa* is confined to the transition zone of summer and winter rainfall areas of the true summer rainfall area, *D. caespitosa* var. *compacta*, in contrast, is a taxon of the true winter rainfall area of the Northern and Western Cape Provinces (Fig. 23). Topographically, they are widely separated along a line of wetter escarpments, the Roggeveld and Hex-River Mts, without any hybrid zone being clearly indicated. However, the occurrence of a few populations of *D. caespitosa* var. *compacta* in the Great Karoo around 33°S and 22°30′E receiving both summer and winter rainfall, is a little mysterious. These plants

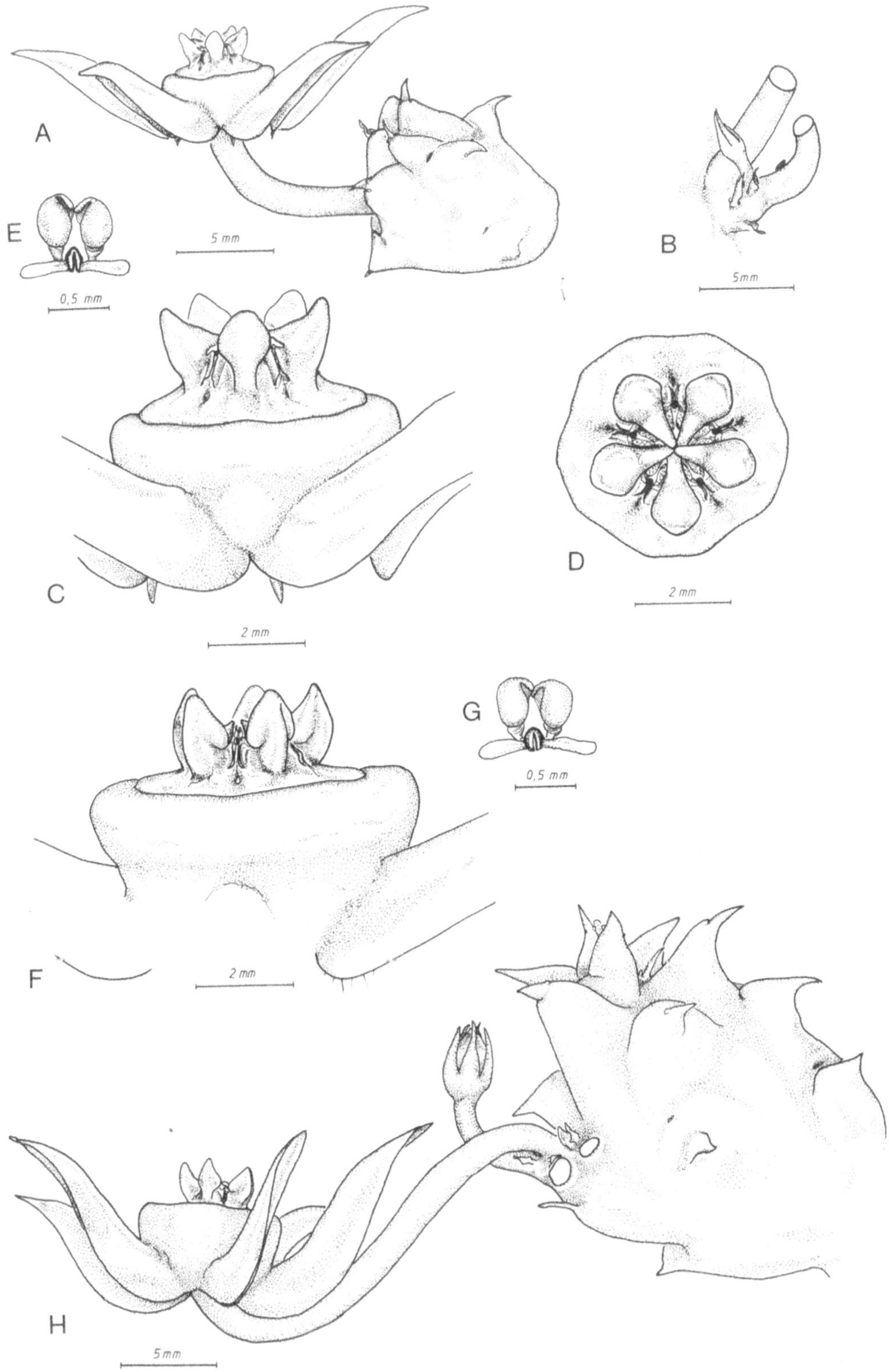

Fig. 24. *Duvalia caespitosa* var. *compacta*. *A* Flowering stem; *B* inflorescence; *C* centre of flower; *D* pollination apparatus in top view; *E* pollinarium; *F* centre of flower; *G* pollinarium; *H* flowering stem. (*A–E* MEVE 245; *F*, *G* MEVE 314; *H* MEVE 252)

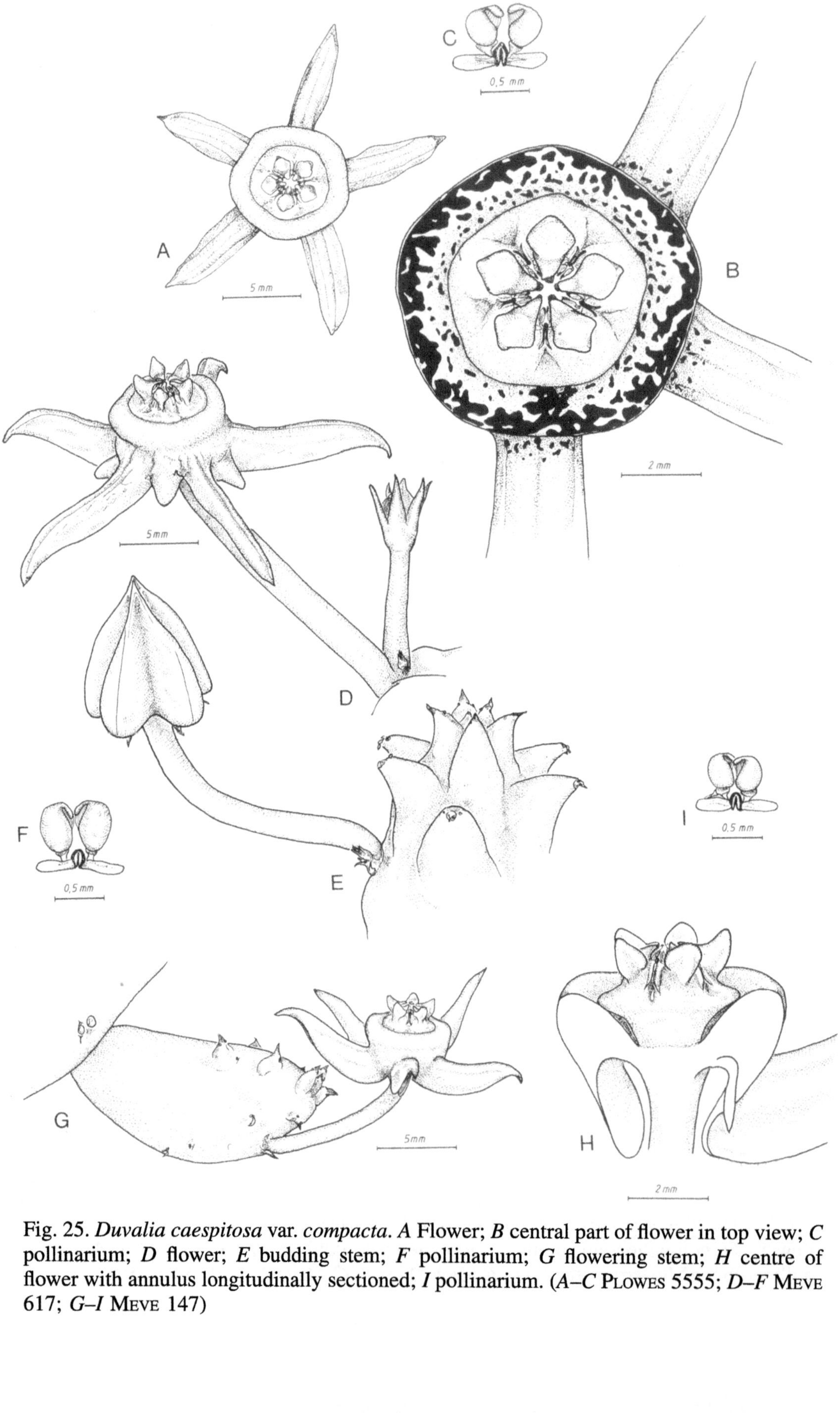

Fig. 25. *Duvalia caespitosa* var. *compacta*. *A* Flower; *B* central part of flower in top view; *C* pollinarium; *D* flower; *E* budding stem; *F* pollinarium; *G* flowering stem; *H* centre of flower with annulus longitudinally sectioned; *I* pollinarium. (*A–C* PLOWES 5555; *D–F* MEVE 617; *G–I* MEVE 147)

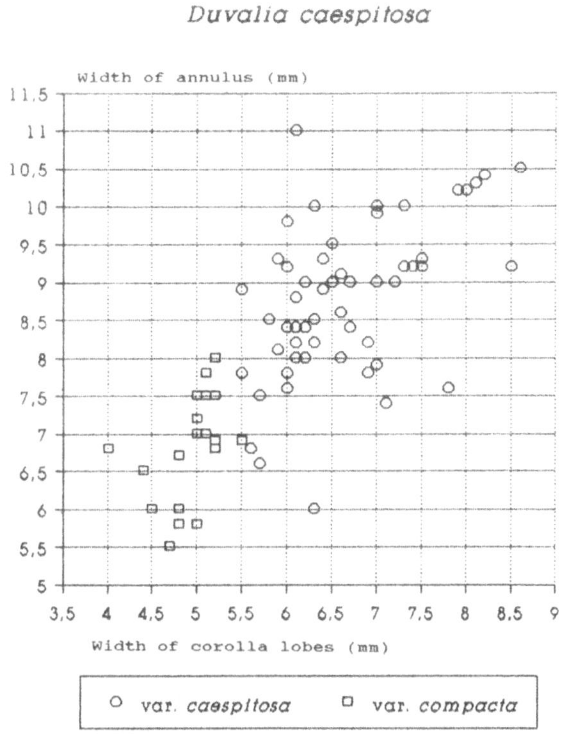

Fig. 26. Scatter diagramm comparing dimensions of annulus and corolla lobes in both varieties of *Duvalia caespitosa*

clearly represent var. *compacta* in all characters of vegetative and floral morphology. Both taxa occur in a wide range of habitats (see descriptions), although var. *compacta* seems to be more tolerant of really dry habitats such as Succulent Karoo. Possibly, the Great Karoo populations of var. *compacta* reflect this tolerance.

The epithet '*compacta*' correctly refers to the mostly compact stems of most of the plants seen. The two varieties are easily separable by the smaller flowers of var. *compacta*, showing a slender annulus and smaller corolla lobes in contrast to the more robust flowers of *D. caespitosa* var. *caespitosa* (Fig. 26). Also, the sparse indumentum and the total lack of clavate hairs along the margins of the corolla lobes as well as the position of the inflorescences at the middle or near the apex of the stems identify var. *compacta*.

D. caespitosa var. *caespitosa* represents the most variable taxon of the whole genus. The sizes of the flowers, their coloration and indumentum vary strongly from region to region (Figs. 21, 22). Even within a population it is rare to find two individual plants with exactly matching flowers. However, a clinal variation concerning the intensity of the indumentum is observable, starting with usually sparsely hairy flowers in the west and finishing with the most hairy ones in the east. Also, in the eastern distribution area vibratile clavate hairs are usually found (Fig. 22). Yellow staminal coronas, in contrast, are restricted to populations of the Little Karoo, demonstrating locally developed forms.

Duvalia pubescens and *D. vestita* constitute the closest relatives of *D. caespitosa*.

Specimens examined

South Africa:

3018 CA: Taaiboshoek, Langberg, Bayer 2778 (NBG!).
3018 CD: 15 mi NE Bitterfontein, Stayner s.n. sub KG 342A/60 (NBG!).
3019 CB: 1 km E Loerieskop Rd to Brakfontein, Liede 2513* (MSUN)
3019 CD: Loeriesfontein, Skietbaan, Bruyns 1504b (NBG!);.
3118 AB: Nuwerus, Nordenstam 2859 (NBG!).
3118 AC: N Landplaas – SW Bitterfontein, Albers & al. K 1331* (MSUN).
3118 BD: 27 km NE Vanrhynsdorp, 3 km E Grootdrif, Plowes 3381 (SRGH).
3118 CA: Papendorp, Olifants River Mouth, Hall 1240 (NBG!).
3118 DA: Vanrhynsdorp, Brauns s.n. (STEU 5776, STEU); Vanrhynsdorp, Herre s.n. (STEU); 15 km S Vanrhynsdorp, Meve 147*, 148* (MSUN).
3119 AA: ca. 11 km NE Farm Grasberg, Theunisdrift, Wisura 2904 (NBG!).
3119 AB: 40 km N Nieuwoudtville, Kokerboomkop, Meve 245*, 247*, 248* (MSUN); Calvinia, Hantam River, s. coll. s.n. sub KG 163/61 (NBG!).
3119 AC: Calvinia, Rietfontein, Bayer 1864 (NBG!); Vanrhyns Pass, Hall s.n. sub NBG 1402/48 (NBG!); 12 km N Nieuwouldtville, Meve 250* (MSUN); ca. 13 km N Nieuwoudtville,"Uithoek", Meve 252* (MSUN); N Nieuwoudtville, Brandkop, Bayer 666 (NBG!); E Vanrhynsdorp, at base of VanRhyns Pass, Meve 261*, 263* (MSUN).
3119 BB: 20 km N Calvinia, Hardy 2989 (PRE!); 50 km NO Calvinia, Plowes 5555 (SRGH!).
3119 BC: 10 km E Clanwilliam/Nieuwoudtville turnoff, Bayer 662 (NBG!); 12 mi NW Calvinia, Bayer 758 (NBG!); 34 km NW Calvinia, Bayer 1866 (NBG!); 40 km NE Calvinia, Plowes 5556 (SRGH); Calvinia, Toren Rd, Stayner s.n. sub KG 162/61 (NBG!).
3119 BD: 5 mi E Calvinia, Bruyns 107/75 (NBG!).
3119 CB: NNE Clanwilliam, Meve & Liede 618* (MSUN).
3119 CD: N Clanwilliam, Botterkloof, Bayer s.n. sub KG 357/75, (NBG!); N Clanwilliam, Botterkloof, Bayer 633, 661, 663 (NBG!); Ceres Karoo, Foot of Botterkloof Pass, Wisura 721 (NBG!).
3119 DA: 5.5 km S Calvinia, Meve 266*, 267 (MSUN); 16 mi S Calvinia, Hall 998 (NBG!).
3119 DB: 10 mi SE of Downes, Hall s.n. sub NBG 265/55 (NBG!).
3120 CA: Kl. Tafelberg, Bruyns 4296 (BOL).
3218 BB: 10 km N Clanwilliam, Bayer 654 (NBG!).
3219 BA: 10 km E Doornrivier, Tanqua Karoo, Lavranos 11735 (SRGH).
3219 CB: Kromrivier between Ceres and Algeria, Bayer 742 (NBG!); Nuwerus, Bruyns 3056 (BOL).
3220 AA: Bottom of Gannaga Pass, Bayer 1931 (NBG!).
3220 AB: Sutherland, 20 km S Middlepos, Lavranos & Pehlemann 18108* (MSUN).
3220 CC: Kookfonteinberg, Bayer 2480 (NBG!).
3220 DB: 82 km N Laingsburg, Moordenaarskloof, Bayer 2135 (NBG!).
3222 DC: S Beaufort West, Wolvekraal (32°58′S 22°44′E), Albers & al. K 1131* (MSUN).
3222 DD: ca. 74 km E Prince Albert, Meve 314*, 315 (MSUN); W Rietbron, Farm Prutkraal, Snyman s.n. sub KG 249/73 (NBG!).
3319 BA: 3 km NE Karoopoort, Stayner s.n. sub KG 496/71 (NBG!).
3320 BA: Baviaans, W of station, Bayer 2111 (NBG!).
3320 BC: Anysberg Pass, Bayer 668 (NBG!).
3321 AD: 2 mi W Floriskraal Dam, Bayer 762 (NBG!).
3322 AB: 31 km E Prince Albert, Meve 310* (MSUN).

3. ***Duvalia corderoyi*** **(Hook. f.) N.E. Br.**, Bot. Mag., sub t. 6245 (1876). Basionym: *Stapelia corderoyi* Hook. f., Bot. Mag., t. 6082 (1874).

Type: Cape Province, Aliwal North Div., 1841, Burke s.n. (holo-, K!).

Description. Stems 12–30 (−45) mm long, 12–25 mm diam., 6-angled, spheroid to ovoid-cylindrical, dark green, glabrous, leaf rudiments ca. 1 mm long, 1 mm wide at the base, stipular glands present. Inflorescence bostrychoid, 1–4 flowers developing successively near the base of the stems, with intense excrement odour; pedicels 15–25 mm long, ca. 2 mm diam.; sepals ca. 4 mm long, ca. 1 mm wide at the base. Corolla 30–45 mm diam., bright reddish-brown to dark purple, rarely bright greenish-brown, lobes longitudinally furrowed, 12–16 mm long, basally ca. 6 mm high, replicate without touching at the margins, margins with purplish, vibratile clavate hairs, 2–4 mm long for up to 2/3 of their length, basal sinuses of the lobes with simple hairs, 300–600 µm long hairs; annulus fleshy, circular to slightly pentagonal, 2.5–4 mm high, 10–15 mm diam., upper rim flattened, upper half of annulus or just the upper rim brownish and maculated in the colour of the corolla, whole annulus densely hairy, hairs simple, ca. 3 mm long, whitish to purple. Coronal disc bright reddish-brown to dark purplish, 6.5–9 mm diam., circular, pentagonal or decagonal, staminal corona lobes 2–3 mm long in top view, dorsal lobules 2–3 mm long, cream to reddish-brown. Pollinia dark reddish-brown, 450–550 µm long, 300–350 µm wide, pellucid margin ca. 250 µm long; caudicles orange, ca. 150 µm long, translator wings ca. 500 µm long; corpusculum ca. 270 µm long. Follicles 10–15 cm long; seeds 5–6 mm long, 3–3.5 mm diam., wing ca. 0.7 mm wide, coma ca. 12 mm long. Chromosome number: $2n = 66$ (vouchers: see * "Specimens examined") (Fig. 27).

Distribution. In central South Africa; mainly in the Great and the Upper Karoo; Free state 29°10′–32°25′ S and 22°35′–26°40′ E (Fig. 28).

Habitat. Mostly associated with Acocks Veld Types 27 and 36, Central Upper Karoo and False Upper Karoo; alt.: 1050–1600 m a. s. l.

Duvalia corderoyi is endemic to the Central and the Upper Karoo, where it favours habitats around an altitude of 1300 m. Summer rainfall is prevalent, only at the southern border of the distribution area, rain distributed over the whole year are to be expected. Apart from the conspicuously six-angled, nearly globose stems, this species with its large flowers, long-haired annulus and the longitudinally furrowed corolla lobes (Fig. 27) recalls one of *D. sulcata*. All types of the corolla indumentum including the extraordinarily papillate or hairy sinuses of the corolla lobes are present in *D. sulcata* subsp. *sulcata* as well. Slightly papillate sinuses are otherwise only known from *D. angustiloba, D. gracilis, D. maculata* and *D. modesta*. The slightly upwards vaulted translator wings, however, are exclusively found in *D. corderoyi* (Fig. 27B, D).

Specimens examined

South Africa:

2926 AA: Bloemfontein, Rayton Ridge, Chaplin s.n. sub OFS 31/76 (NBG!), Bloemfontein, Potts 2586 (K!); s.coll. s.n. OFS Bot.Garden 268/79 sub K 1000*; OFS, Bloemfontein, ex hort. NBG 125/69 *(MSUN); Winter's Valley, Wisura s.n. (NBG!)

Plate 1. *A Duvalia angustiloba*, MEVE 321, ca. 2.5.x; *B D. caespitosa* var. *caespitosa*, MEVE 454, ca. 1.5x; *C D. caespitosa* var. *caespitosa*, MEVE 283, ca. 2.5x; *D D. caespitosa* var. *compacta*, MEVE 147, ca. 4x; *E D. corderoyi*, MEVE 351, ca. 1.5x; *F D. elegans*, MEVE 395, ca. 2.5x

Plate 2. *A Duvalia gracilis*, MEVE 460, in habitat, 62 km W Cradock, ca. 2x; *B D. immaculata*, MEVE 399, ca. 3x; *C D. maculata*, MEVE 353, ca. 3x; *D D. modesta*, MEVE 332 (left) & MEVE 334 (right), ca. 2x; *E D. pillansii*, ex hort. Karoo Garden, ca. 2.5x; *F D. parviflora*, BAYER s.n. sub KG 49/73, ca. 3x

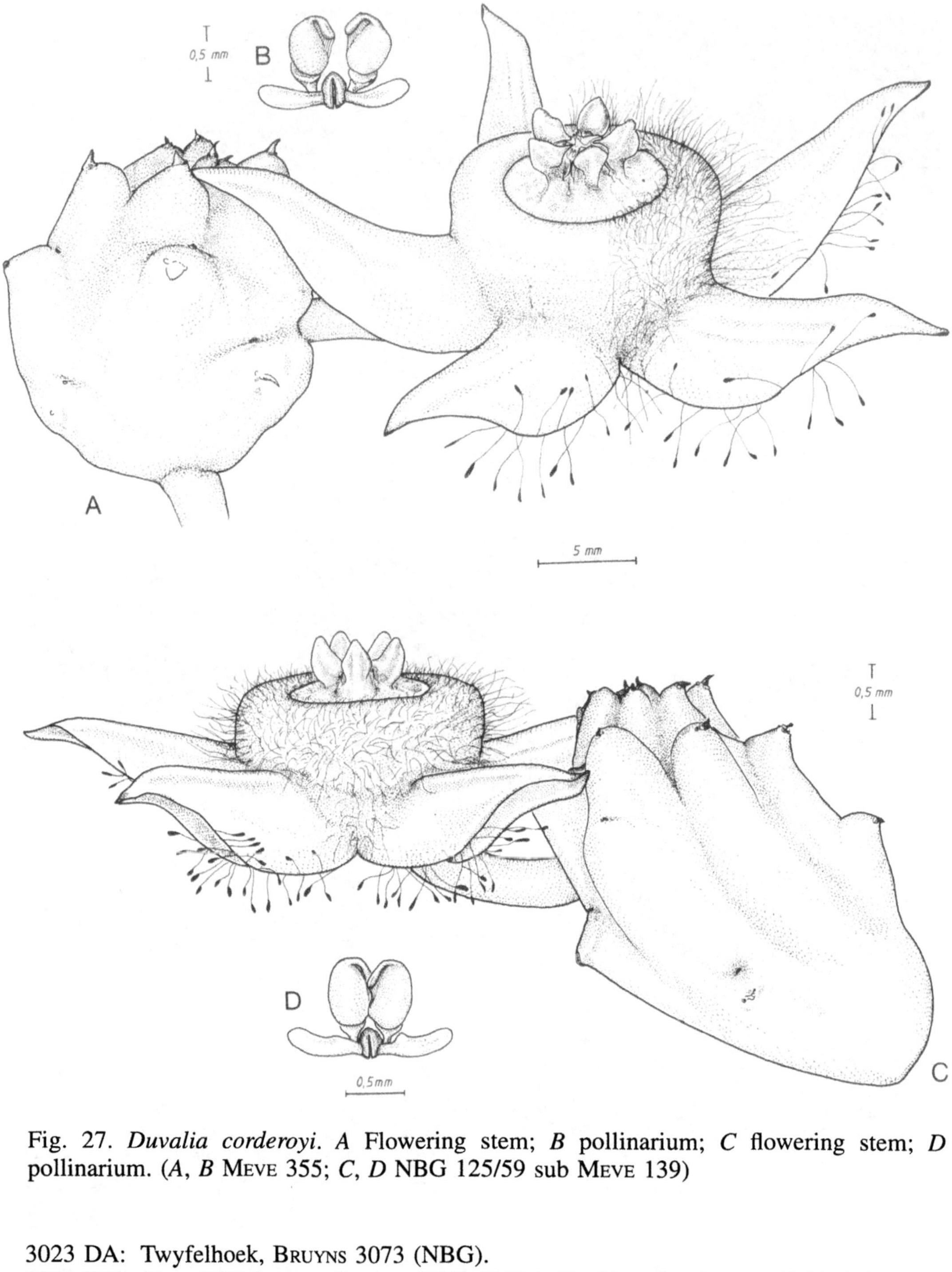

Fig. 27. *Duvalia corderoyi*. *A* Flowering stem; *B* pollinarium; *C* flowering stem; *D* pollinarium. (*A*, *B* Meve 355; *C*, *D* NBG 125/59 sub Meve 139)

3023 DA: Twyfelhoek, Bruyns 3073 (NBG).
3023 DD: 4 km S Vlokpoort, Bruyns 3076 (BOL); Deelfontein, Bruyns 4223 (BOL).
3024 AD: Philipstown, Markötter s.n. sub STEU 5507 (STEU).
3024 CA: De Aar, Bruyns 1923*, s.n. sub K 1655* (NBG!).
3024 CD: Palmietfontein, Bruyns 1620 (NBG!).
3122 DD: ca. 25 km W Kromrivier, Meve 351*, 352*, 354* (MSUN); ca. 24 km W Kromrivier, northern slope of Aasvoelberg, Meve 355* (MSUN).

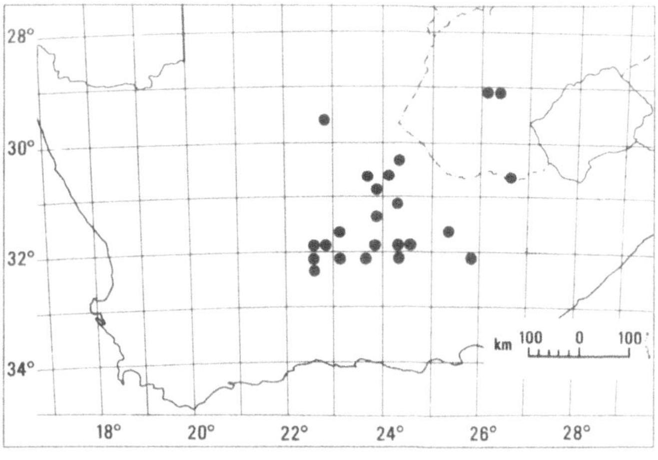

Fig. 28. Distribution of *Duvalia corderoyi*

3123 BD: Richmond, Bruyns 3072 (NBG).
3123 DD: 1 km E Murraysburg, Bayer 2390a (NBG!).
3123 CA: Biesiespoort, Bruyns 4229 (BOL).
3124 CD: 7 km Krugerskraal, Sneeuberg, Branch 318 (NBG!).
3124 DC: New Bethesda, Aasvoelkrans, Bayer 2357* (NBG!).
3125 CB: Middleburg Div., near Conway, Pillans 191 (K!).
3222 BA: ca. 13 km N Beaufort West, Nuwerus, Bayer 632 (NBG!, SRGH).
3222 BC: Stolshoek, Bruyns 3376 (BOL).
3223 AA: E Nelspoort, Kruidfontein, Bayer 2395 (NBG!).
3223 BA: Rietfontein, Bruyns 4240 (BOL).
3224 AB: N Graaff Reinet, St. Olive Ouberg, Bayer 2351 (NBG!); Graaff Reinet, top of Ouberg Pass, Roussouw 655 (NBG!).
3225 BB: 9 km Lake Arthur from Spioenkop, Bruyns 1779a (NBG!).

4. ***Duvalia elegans*** **(Masson) Haw.**, Syn. Pl. Succ. 44 (1812). Basionym: *Stapelia elegans* Masson, Stap. Nov., 19, t. 27 (1796).

Type: Icon. Masson, Stap. Nov., t. 27 [lecto-, designated by N.E. Brown (1908)].

= *Stapelia radiata* Jacq., Stap., t. 12 (1806–1819) [nec Sims 1803]. Type: Icon. Jacq., Stap., t. 12 [lecto-, designated by N.E. Brown (1908)].

= *D. jacquiniana* (Schult.) Sweet, Hort. Brit. 1st ed, 276 (1827); Loud., Encycl. Pl., 202, (1841) as *D. jacquini*. Basionym: *Stapelia jacquiniana* Schult., in H. Roemer & Schult, Syst. Veg. **6**: 45 (1820). Type species: Icon. Jacq., Stap., t. 12 (lecto-, designated here).

= *D. elegans* var. *seminuda* N.E. Br., in Thiselton-Dyer, Fl. Cap. **4**(1): 1028 (1908). Type species: Southwestern Cape Province, 1/2 mile west of Riversdale, Pillans 682 (holo-, BOL).

= *D. elegans* var. *elegans* f. *magnicorona* A.C. White & B. Sloane, Stap. **2**: 741 et Stap. **3**: 1144 (1937). Type species: Icon. White & Sloane, Stap. 2, 741, Fig. 729 (based on White & Sloane-No 178-1, s. loc., see p. 745) (lecto-, designated here).

Description. Stems 20–60 mm long, 8–18 mm diam., 4–5-angled, ovoid-cylindrical, rather rounded due to the only slightly raised podaria, dark green, glabrous, leaf rudiments 1–1.5 mm long occasionally with glandular hairs along their upper margins, stipular glands minute, rarely absent. Inflorescence bostrychoid, 1–5 flowers developing successively from the stem base, with intense carrion stench. Pedicels 12–25 mm long, ca. 1 mm diam.; sepals ca. 3 mm long, 1 mm wide at the base. Corolla 15–22 mm diam., adaxially blackish-purple, brilliant, covered with simple, slightly bent, dark purple hairs 1–3 mm long, lobes 6.5–10 mm long, 3.5–6 mm wide at the base, somewhat spreading, margins slightly replicate, apex of corolla lobes sometimes glabrous; annulus indistinctly developed, circular, 0.8–1.5 mm high, 6–8 mm diam., densely hairy, hairs 1–2 mm long. Coronal disc reddish-brown to dark purple, 5–7 mm diam., circular, pentagonal to decagonal, sometimes somewhat undulate, touching the rim of the annulus or rather overlapping it, dorsal lobules of the staminal corona reddish-brown, 1.2–1.8 mm long, with horizontally to vertically orientated apices, acute to obtuse. Pollinia orange-yellow to reddish-brown, 280–340 μm long, ca. 200 μm wide, pellucid margin ca. 130 μm long; caudicles orange, ca. 100 μm long, translator wings ca. 300 μm long; corpusculum ca. 130 μm long. Follicles 8–13 cm long; seeds ca. 5 mm long, ca. 3 mm diam., wing ca. 0.5 mm wide, coma ca. 15 mm long. Chromosome number: $2n = 22$ (vouchers: see * in "Specimens examined"). (Figs. 29, 30).

Distribution. Southern Western Cape: Little Karoo, Worcester-Robertson Karoo; 33°35′–34°25′ S and 18°20′–21°30′ E (Fig. 31). Namibia?

Habitat. Element of the Acocks Veld Type 26, Karroid Broken Veld; alt.: 150–300 m a. s. l.

The plant depicted by Masson is identical with the one illustrated in the Botanical Magazine (Sims 1809) and used by Jacquin (1806–1819) for the description of his "*Stapelia radiata*". It is reported that Jacquin received his plant under this name from England and therefore used the epithet *radiata* in his description. However, he obviously was not aware at this time that Sims (1803) had already described a plant of the *D. caespitosa* alliance under that name. Instead,

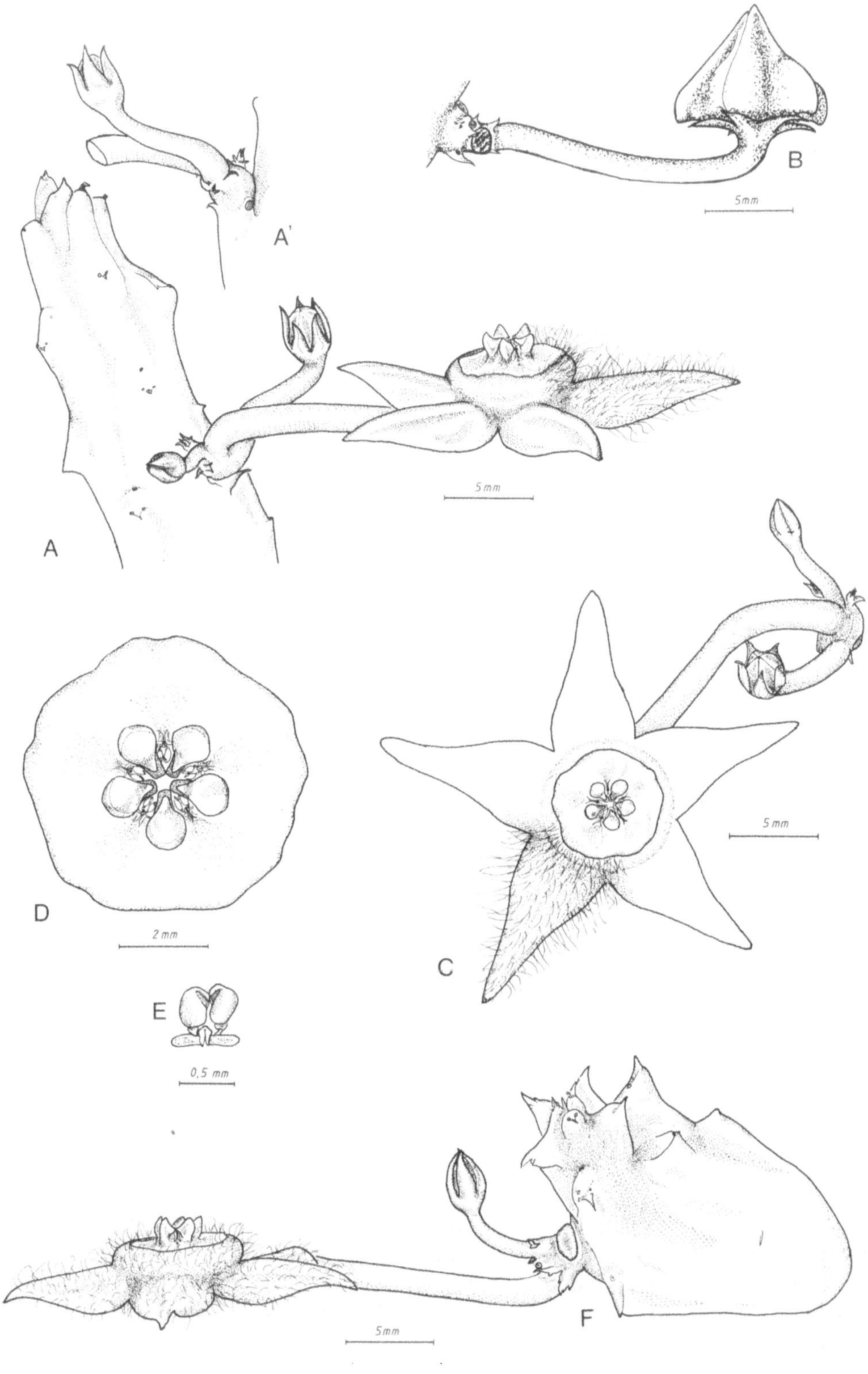

Fig. 29. *Duvalia elegans*. *A*, *A'* Inflorescence; *B* bud; *C* inflorescence in top view; *D* pollination apparatus in top view; *E* pollinarium; *F* flowering stem in lateral view. (*A–E* MEVE 395; *F* ALBERS & al. sub K 1381)

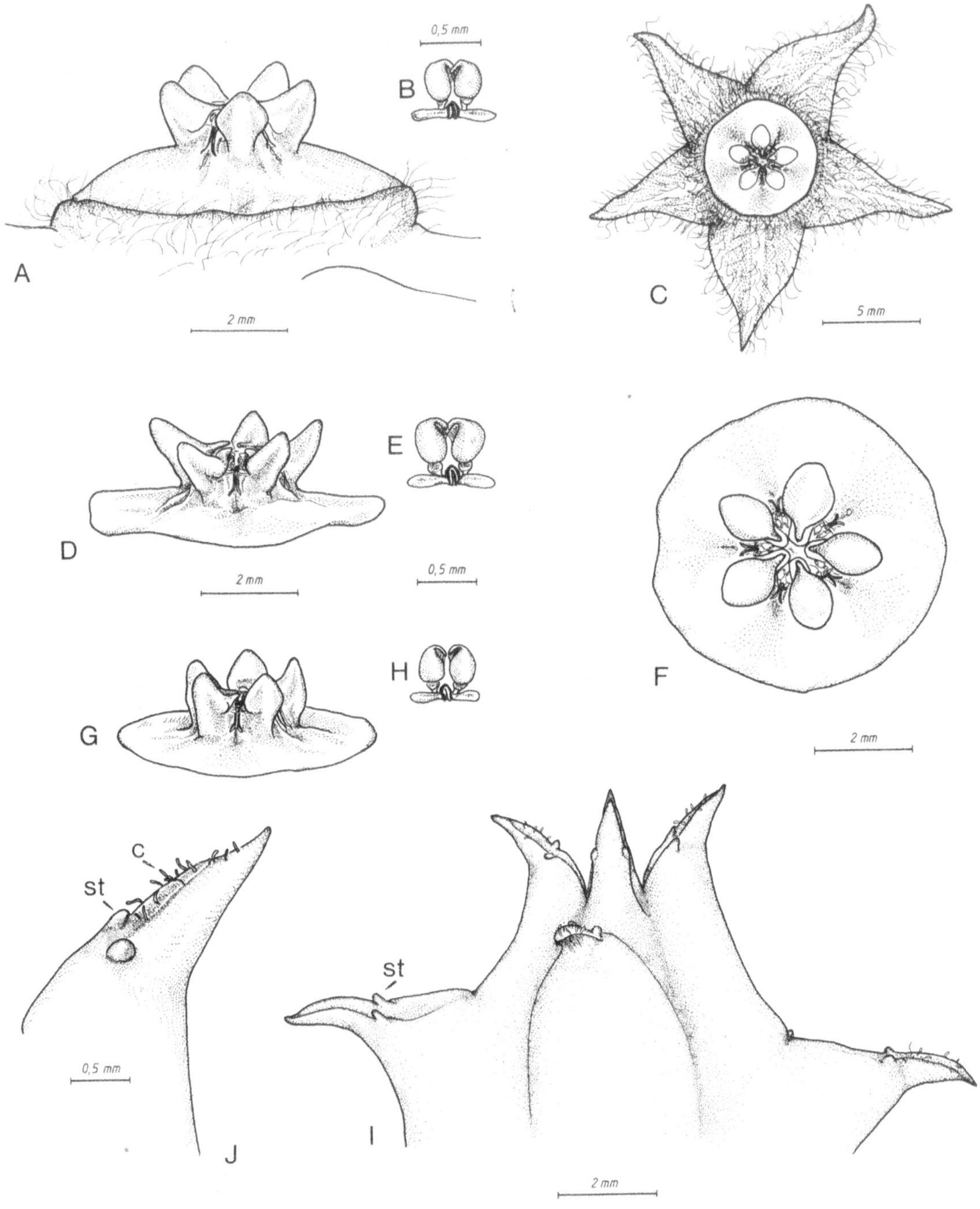

Fig. 30. *Duvalia elegans*. *A* Central part of flower in lateral view; *B* pollinarium; *C* flower in top view; *D* pollination apparatus; *E* pollinarium; *F* pollination apparatus in top view; *G* pollination apparatus; *H* pollinarium; *I* apex of young stem; *J* podarium with stipular glands (*st*) and colleters (c). (*A*, *B* MEVE 383; *C–F* LEACH & FORRESTER 16844; *G*, *H* ALBERS & al.sub K 1381; *I*, *J* MEVE 393)

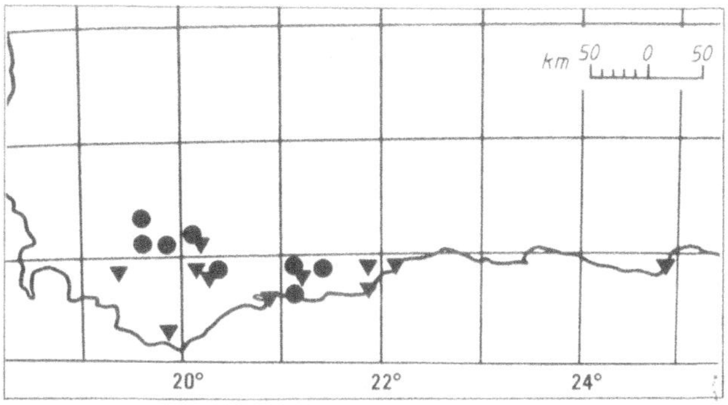

Fig. 31. Distribution of *Duvalia elegans* (●) and *D. vestita* (▼)

JACQUIN unfortunately described and illustrated the "true" *D. radiata* SIMS as *D. replicata* (cf. BROWN 1908).

SCHULTES (1820) recognized JACQUIN's error only with respect to one of two taxa concerned, and he established the name *S. jacquiniana* for *S. radiata* JACQ., thus creating another synonym for *D. elegans*. LOUDON (1841) finally repeated this "half correction" and named *S. radiata* JACQ. *D. jacquini* LOUD.

Duvalia elegans (Figs. 29, 30) is characteristic of the Karroid Broken Veld of the Worcester-Robertson Karoo, though additional localities near Montagu and Riversdale are known (Fig. 31). Usually rather humid localities, always around 200 m a. s. l., are preferred.

It remains unresolved whether the Namibian collection of EBERLANTZ illustrated by W. GIESS (WIND, and pers. comm.) originates from a wild population (compare also HUBER 1967, PLOWES 1982). Even though no original material is preserved, the drawings of W. GIESS undoubtedly show *D. elegans*, a species never found again in Namibia. Considering the scattered distribution of *D. maculata*, relict populations of *D. elegans* in Namibia seem imaginable. The area around Aus is a well-known exclave of the "Succulent Karoo Region" (JÜRGENS 1991). Furthermore, comparable disjunct distribution areas are frequent in the Cape floristic region as a whole concerning succulent species (e.g., *Quaqua mammillaris*, *Pelargonium carnosum s. l.*) as well as non-succulent ones (*Nymania capensis*, *Aitonaceae*, *Protea sulphurea*, *Proteaceae*).

The type species of the genus possesses untypical features regarding floral morphology (flat annulus, large coronal disc), However, the genome as well as the small and central distribution area point to a basal position of *D. elegans*. Its distribution area is the westernmost of the diploid species. The strongly expanded coronal disc represents a phylogenetically more recent development. Therefore, *D. elegans* should be understood as the most advanced species within the basal, diploid group.

Duvalia elegans differs from its nearest relative, *D. vestita* (Fig. 45), not only in the size of the genome but also by the spreading corolla lobes with longer hairs on nearly the whole adaxial corolla surface. The superficially similar *D. pubescens* and *D. caespitosa* p.p. are more distantly related.

Specimens examined

Namibia:

2615 CA: Lüderitz (or Aus ?), GIESS s.n. (WIND, pencil scetch only).

South Africa:

3319 DA: 18 mi E Worcester, HALL 2012 (NBG).

3319 DC: Robertson, Noddee, BAYER 1414 (NBG!); Langvlei (Niesenbarend) Noree, BAYER s.n. sub KG 343/71 (NBG!); 3 km W Langvlei, ROUSSOUW 19 (NBG!);

3319 DD: Langvlei, ca. 10 km NW Robertson, MEVE 379* (MSUN); 3 mi W Robertson, HALL 873 (NBG!).

3319 DD: Between Langvlei and Voorspoed, ALBERS & al. K 1380* (MSUN); 7.5 km N McGregor, MEVE 382*, 383* (MSUN); Robertson, PLOWES 3080* (MSUN).

3320 CC: S Bonnievale, Bokkraal, BAYER 653 (NBG!); Drew, BAYER 764 (NBG!); Ashton, BAYER 1932 (NBG!); Touwsrivier turnoff on Kingna River, BRUYNS 2206 (NBG!); Montagu Distr., Cayman-Kloof, LITTLEWOOD s.n. sub Kew 30754 (K!); Bonnievale, MARLOTH 6499 (PRE!); 14 km NW Stormsvlei, MEVE 393*, 394*, 395* (MSUN); near Ashton, PILLANS 75 (K!); Montagu, W end of Kogmans Kloof, s. coll. s.n. sub KG 393/61 (NBG!).

3420 AB: Stormsvlei, FASCIO 14 (NBG!).

3421 AA: S Riversdale, LEACH & FORRESTER 16844* (NBG!).

3421 AC: 20 mi SW Riversdale, KRAMER s.n. (PRE!).

5. ***Duvalia gracilis*** **Meve, spec. nova**

Diagnosis. Plantae tegetes densas ex caulibus 10–35 mm longis formantes; rudimenta foliaria 3–4 mm longa, peracuta; pedicellus 10–20 mm longus; gemmae florales peracutae, conicae; flores 16–22 mm diametro, superficie interiore rubrofusca vel fusco-purpurea praediti; annulus pentagonus, 6–7 mm diametro, 0.6–0.8 mm altus, marginem complanatam formans, zona centrali viridiuscula vel ex parte viridiusculo-maculata ornatus, papillis trichomatoideis usque 200 μm longis (ad modum pars tertia basalis loborum liberorum) vestitus; corona viridi-flava vel viridi-fusca, disco coronali ca. 4 mm diametro praedita, lobos coronarum staminales ca. 1.5 mm latos evolvens; pollinium flavum; fructus 10–16 cm longi; semina 4–5 mm longa, plus quam 3 mm lata, margine integra praedita.

Type: Eastern Cape 1 km E Farm Krugerskraal, 18. Nov 1986, Meve 346 (holo-, K; iso-, NBG, MSUN).

Description. Stems 10–35 mm long, 7–20 mm diam., 4–5-angled, ovoid to cylindrical, dark green to grey-green, glabrous, leaf rudiments 2.5–4 mm long, very acute, stipular glands small. Inflorescence bostrychoid, 1–4 prostrate flowers developing successively from near the stem base, with slight excrement odour. Pedicel: 10–20 mm long, ca. 1 mm diam.; sepals 3–4 mm long, ca. 1.2 mm wide at the base, green, glabrous, acute. Corolla 16–22 mm diam., reddish-brown with greenish tinge or brown, free corolla lobes 7–9 mm long, 4–5 mm diam., almost totally replicate but margins not touching each other, basal third, at least up to the middle, covered with hair papillae up to 200 μm long; annulus 6–7 mm diam., 0.6–0.8 mm high, pentagonal, flattened with slight depressions by the corolla lobes, adaxial surface of central annulus and the basal regions of the corolla lobes green or maculated with greenish-brown, covered with hair papillae 50–200 μm long. Corona greenish-yellow or greenish-brown with reddish-brown markings along the guide rails, coronal disc ca. 4 mm diam., circular, pentagonal or slightly decagonal, lobes of staminal corona ca. 1.5 mm long. Pollinia yellow, ca. 370 μm long, 210 μm wide, pellucid margin 120–140 μm long; caudicles orange-coloured, ca. 100 μm long, translator wings ca. 350 μm long; corpusculum 180 μm long. Follicles 10–16 cm long; seeds ca. 4.5 mm long, 3 mm in diam, wing 500–600 μm wide, coma 10–12 mm long. Chromosome number: $2n = 22$ (vouchers: see * in "Specimens examined"). (Fig. 32).

Distribution. Eastern Cape, 31°45′–32°00′ S and 24°15–25°15′ E (Fig. 20).

Habitat. In karroid formations, mostly of disturbed types; alt.: 1300–1600 m a. s. l.

Duvalia gracilis was collected by W.R. Branch in 1984 for the first time. It was initially identified as *D. maculata* (NBG, *in schedis*) due to its slightly maculated annulus. The size and extent of the trichomes of the adaxial corolla surface of *D. gracilis* is characteristic (Figs. 8A, 32C, F). The pollinia are particularly large in comparison to other diploid and small-flowered species. With its flattened annulus and the greenish-brown, rather than yellow corona, *D. gracilis* can be easily distinguished from *D. maculata*. Also, *D. gracilis* is restricted to comparatively high altitudes between 1300–1600 m, in which *D. angustiloba* never occurs and *D. maculata* is only rarely found. Here, it is typically confined to overgrazed, disturbed habitats, often dominated by asteraceous shrubs.

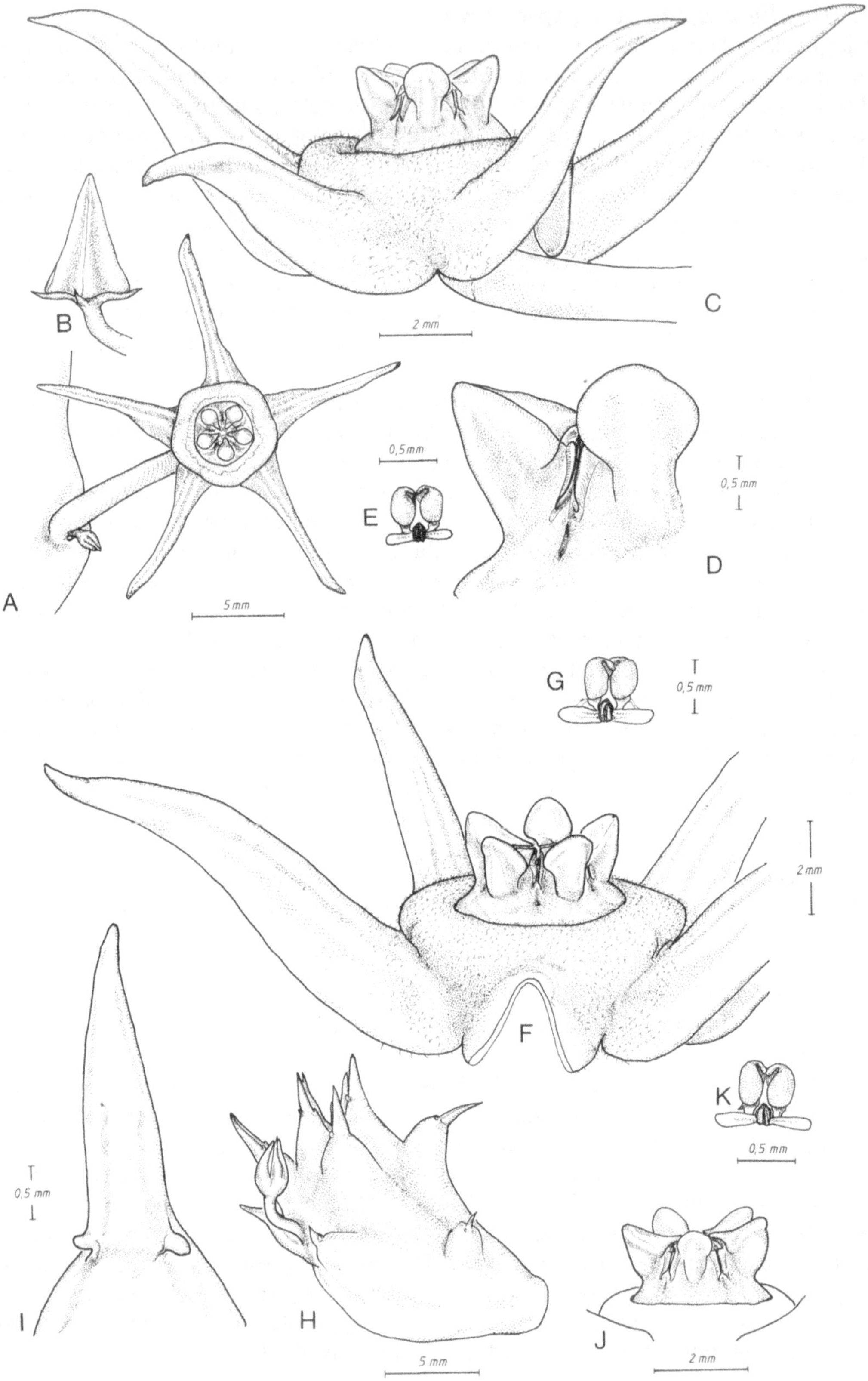
A
B
C
D
E
F
G
H
I
J
K
2 mm
5 mm
0,5 mm
0,5 mm
0,5 mm
2 mm
0,5 mm
0,5 mm
5 mm
2 mm

It should be considered whether *D. gracilis* could constitute the result of wild hybridization between two diploid species, e.g., *D. angustiloba* and *D. maculata*, which share a number of similarities in leaf rudiments, corolla and corona morphology with the new species. However, stapeliad hybrids tend to look intermediate between their parents (cf. BARAD 1990, and this paper), so a hybrid of these two species would not be expected to show characters such as the flattened top rim of the annulus, the considerably amplified indumentum and the coloration of the corolla, corona and pollinia displayed by *D. gracilis* but not by the putative parents. An artificial hybrid produced in Münster between *D. angustiloba* (MEVE 321) and *D. maculata* (BAYER 2008) confirmed this suspicion, lacking any of the characters unique to *D. gracilis*. It seems unlikely, therefore, that *D. gracilis* originated in this manner. Supporting the independent status of *D. gracilis*, its current distribution area is allopatric with *D. angustiloba* and *D. maculata*. Because *D. caespitosa* is reported to share localities with *D. gracilis* (P. BRUYNS, pers. comm.), this species must also be taken into account as a potential progenitor. However, *D. caespitosa* is always tetraploid and *D. gracilis* is a diploid. Therefore, *D. caespitosa* cannot serve as a direct possible ancestor for *D. gracilis*.

Specimens examined

South Africa:

3124 AD: E Cyphorwater, BRUYNS 3321 (BOL).

3124 CD: N Graaff Reinet, Krugerskraal, BRANCH s.n. (NBG!); 1 km E Krugerskraal farmhouse, MEVE 345* (MSUN).

3125 CC: 62 km W Cradock, MEVE 460* (MSUN).

Fig. 32. *Duvalia gracilis*. *A* Flower in top view; *B* bud; *C* flower in lateral view; *D* section of gynostegium with two staminal corona lobes in lateral view; *E* pollinarium; *F* flower; *G* pollinarium; *H* stem; *I* podarium with leaf rudiment and pair of stipular glands; *J* pollination apparatus in lateral view; *K* pollinarium. (*A–E* MEVE 345); *F–I* MEVE 346; *J*, *K* MEVE 460)

6. ***Duvalia immaculata*** **(A.C. Lückh.) M.B. Bayer ex L.C. Leach**, S. Afr. J. Bot. **55**: 268 (1989). Basionym: *D. maculata* var. *immaculata* A.C. Lückh., in A.C. White & B. Sloane, Stap. **2**: 768 (1937).

Type: Icon. (Photo) White & Sloane, Stap. **2**: 768, Fig. 769 [lecto-, designated by Leach (1989)].

= *D. immaculata* (A.C. Lückh.) M.B. Bayer, *nomen inval.*, Bull. Afr. Succ. Pl. Soc. **12**: 27 (1977).

Description. Stems 20–50 mm long, 6–12 mm diam., cylindrical to slightly ovoid, procumbent and often rhizomatous, green, glabrous, rudimentary leaves less than 1 mm long, stipular rudiments macroscopically invisible. Inflorescence bostrychoid, 1–3 flowers developing successively, mostly held erect above stem level but sometimes prostrate, with exrement stench. Pedicels 5–25 mm long, 1–1.5 mm diam.; sepals ca. 2 mm long. Corolla 20–30 mm diam., dark chocolate, glabrous with dull surface; lobes 8–12 mm long, acute, spreading, replicate to the base into vertical plates; annulus 8–11 mm diam., obscurely pentagonal, with its sides sloping underneath it, 2–3 mm deep, the rim undulated and rising considerably above the corona. Corona sunken into the annulus, straw-yellow; coronal disc 3–4 mm diam.; staminal corona lobes stout, the inner parts less than 1 mm long, incumbent on the anthers, the outer parts ovoid, ca. 1 mm broad, 1.5 mm long, dorsally obtuse, close to the level of the inner parts. Pollinia yellow, ca. 450 μm long, 200 μm diam., pellucid margin 250–300 μm long, considerably longer than half the length of the pollinium; caudicles bright orange, ca. 100 μm long, translator wings with acute tips, bright orange, ca. 350 μm wide; corpusculum ca. 200 μm longs. Follicles 8–15 cm long; seeds ca. 6 mm long, 3.5 mm diam., wing 0.6–1 mm wide, coma 12–15 mm long. Chromosome number: $2n = 66$ (vouchers: see * in "Specimens examined"). (Fig. 33).

Distribution. Southeastern Western Cape, along a small coastal strip between 20°30′ and 22°15′ S (Fig. 34).

Habitat: ± restricted to the Acocks Veld Type 47, Coastal Fynbos, alt.: 50–150 m a. s. l.

This species was considered a variety within *D. maculata* for more than 50 years. This ranking was justified because the somewhat similar morphology of the annuli was considered significant. However, *D. immaculata* is possibly the most derived species of the genus. The deep and mighty annulus overtopping a gynostegium with blunt, ovoid corona lobes (Fig. 33A, C), the acute translator wings (Fig. 33E), the hexaploid genome ($2n = 66$) and the marginal distribution area demonstrates its considerable distance from the other species. Ecogeographically, *D. immaculata* is also isolated in the Coastal Fynbos. Such an extreme change of habitat from semi-arid areas, where the ancestors appear to have originated, into humid coastal vegetation, is very rare in stapeliads. Only *Orbea variegata* occurs frequently along the coast of the southwestern Cape, but the latter species inhabits karroid situations in addition.

Specimens examined

South Africa:

3420 BC: Breede River, ca. 8 km from sea, West Bank, Stayner sub KG 70/73 (NBG!).

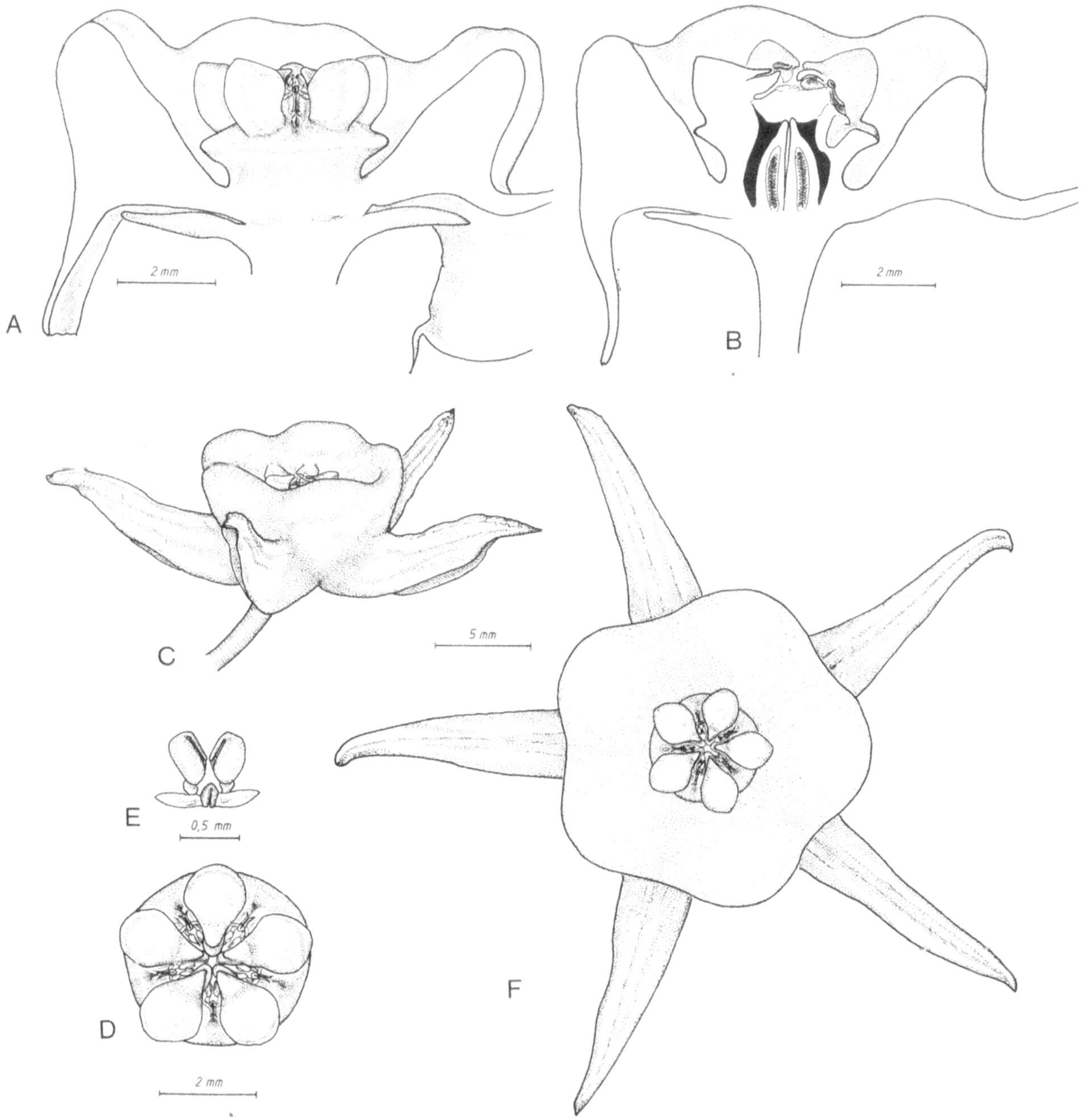

Fig. 33. *Duvalia immaculata. A* Centre of flower with annulus longitudinally sectioned; *B* centre of flower in longitudinal section; *C* flower; *D* pollination apparatus in top view; *E* pollinarium; *F*, flower in top view. (*A–E* MEVE 400; *F* MEVE 399)

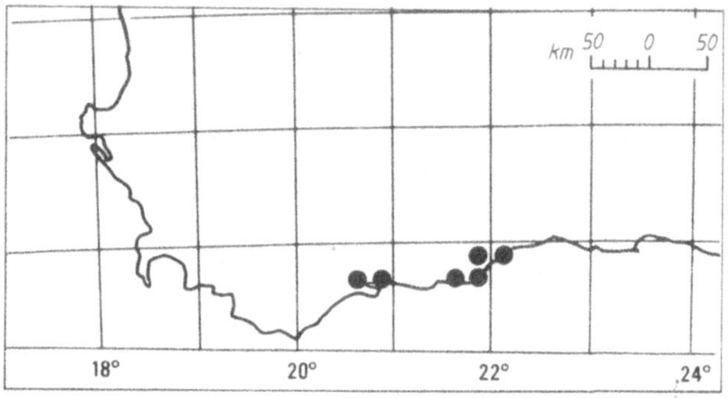

Fig. 34. Distribution of *Duvalia immaculata*

3420 BD: Gouritzmond, Meve 399*, 400*, 401* (MSUN); Infanta, Breede River Mouth, Stayner s.n. (NBG!).
3421 BB: N Cooper Siding, Bayer 743 (NBG!); Henry's Rus (Gouritz Bridge Area to Herbertsdale), Bruyns 1188 (NBG!); Mudlark, Bruyns 4368 (BOL).
3421 BD: Gouritzmond (centre of town), Bayer 15 (NBG!); Gouritz River mouth, Perry 1474 (NBG!).
3422 AA: NW Klein Brak, SW of river, Bayer 2298 sub K 1500* (NBG!).

7. ***Duvalia maculata*** **N.E. Br.**, in Thiselton-Dyer, Fl. Cap., **4**(1): 1033 (1908); A.C. White & B. Sloane, Stap. **2**: 767 (1937) (nec 768, fig. 770).

Type: Eastern Cape, Aberdeen Div., near Aberdeen Road (Grid: 3224 CB), E. Pillans sub Pillans 31 (holo-, K!; iso-, SAM!, BOL).

= *D. minuta* Nel, in A.C. White & B. Sloane, Stap. **3**: 1168 (1937). Type: Namibia: Mickberg (Great Karas Mts., Grid: 2718CB), Icon. (Photo) White & Sloane, Stap. **3**: 1168, fig. 1218 (lecto-, designated here).

Description: Stems 10–40 (−60) mm long, 8–18 mm diam., 4-(5-)angled, ovoid to cylindrical, dark green to grey-green, glabrous, podaria raised, leaf rudiments 2–3 mm long, very acute, stipular glands very small. Inflorescence bostrychoid, 1–8 usually prostrate flowers developing successively from the basal half of the stems, almost without odour. Pedicels 10–25 mm long, ca. 1 mm in diam; sepals ca. 3 mm long, ca. 1 mm wide at the base. Corolla 15–20 mm diam., reddish-brown, lobes 5–7 mm long, 5–6.5 mm wide, ca. 3 mm high at the replicate base, sinuses with a few papillae, basal margins of lobes hairy, trichomes simple, 0.2–2.5 mm long; annulus whitish or cream, variably maculated with reddish-brown dots, obtusely pentagonal, bowl-shaped, basally constricted, with undulated rim considerably overtopping the coronal disc, 5–8 mm diam., 1.5–2 mm high, variably covered with hair papillae 50–250 µm long. Corona canary-yellow, coronal disc 3–3.6 mm diam., fleshy, convex, lobes of staminal corona ca. 1.2 mm long and wide with flattened top. Pollinia yellowish-green, ca. 300 µm long, up to 200 µm wide, pellucid margin 120–150 µm long, hook-shaped; caudicles yellow, ca. 100 µm long, translator wings 350–400 µm long; corpusculum ca. 130 µm long. Follicles 6–12 cm long; seeds 4–5 mm long, less than 3 mm diam., wing up to 500 µm wide, undulate or somewhat serrate, coma ca. 14 mm long. Chromosome number: 2n = 22 (vouchers: see * in "Specimens examined") (Fig. 35).

Distribution. Cape Provinces, Great Karoo, Upper Karoo: 28°50′–32°55′ S and 21°10′–26°00′ E; Namibia: Karas Mountains (Fig. 20).

Habitat. In many different types of karroid vegetation (see below), 700–1500 m a. s. l.

The presentation of a specimen of *D. caespitosa* as *D. maculata* (Phillips 1932) was responsible for long-term confusion. Even B. Bayer, a true expert, doubted the existence of this species for a long time (Bayer 1977, 1984b).

It is noteworthy that *D. maculata* habitats are not homogeneous. This species can be found often in open karroid formations at medium high altitudes (at localities dominated by *Ruschia spinosa*, P. Bruyns, pers. comm.). In addition, *D. maculata* can also be found at an elevation of 1500 m, where it competes against grasses in more humid and cooler conditions. *Duvalia maculata* frequently shares its habitat with *D. corderoyi*.

The maculation of the annulus, which gives the species its name, is found in every specimen without exception. The dimension of the annulus, however, varies considerably. This is also the case for the extent of the corolla indumentum (Fig. 35).

Duvalia maculata is easily recognizable. The intense yellow and compact coronas are as characteristic of the species as are the bright green pollinia. Also, undulated wings of the seeds are unique to this species.

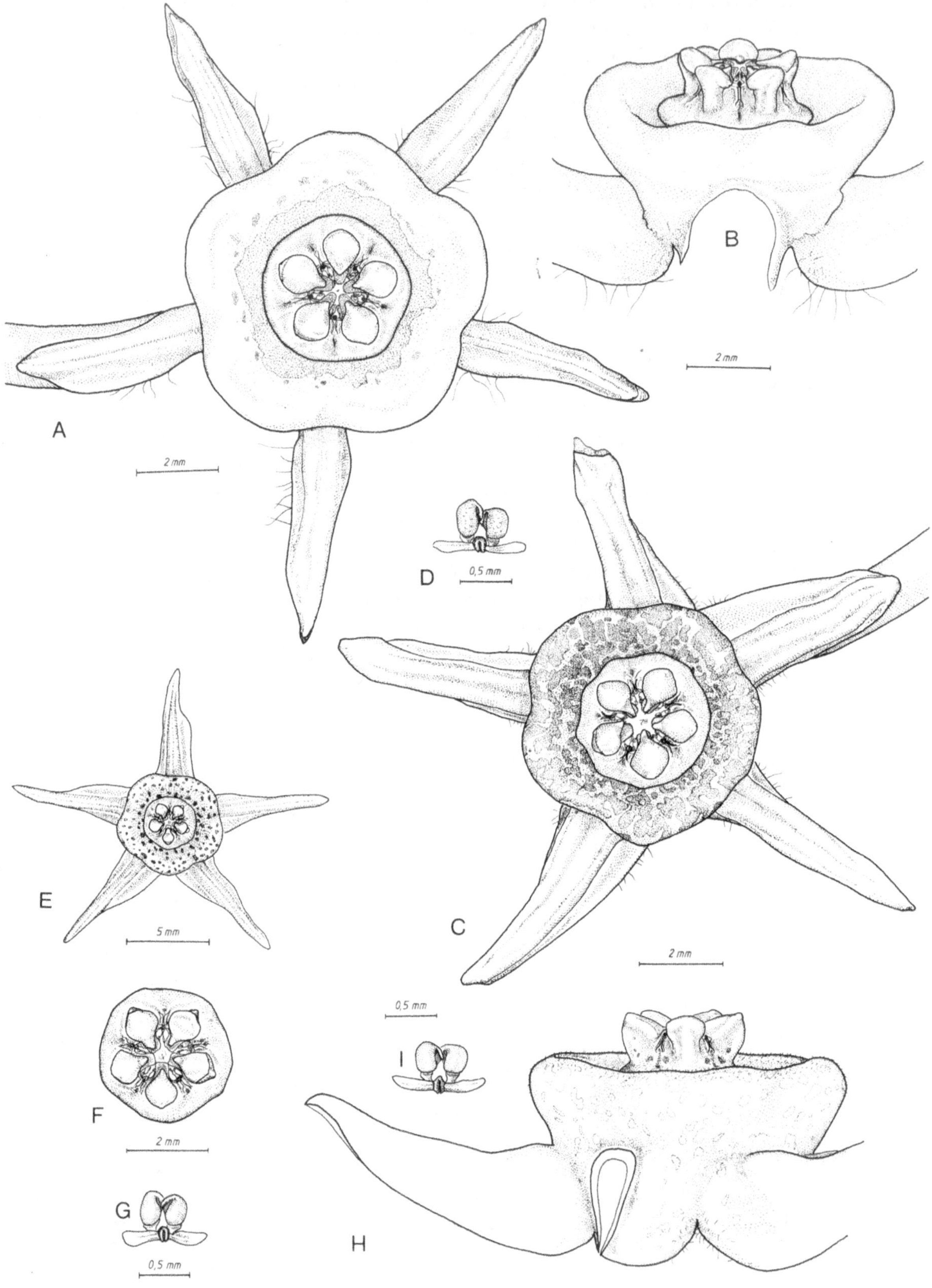
A
2 mm
B
2 mm
D
0,5 mm
E
5 mm
C
2 mm
0,5 mm
I
F
2 mm
G
0,5 mm
H

Duvalia maculata shares the possession of very acute and long leaf rudiments with *D. gracilis*.

Specimens examined

South Africa:
2822 DD: Rudesheim, Bruyns 4505 (BOL).
2921 AC: 15 km SW Kenhardt, Bruyns 4542 (BOL).
2921 CB: 44 km S Kenhardt, Bruyns 3462 (BOL).
2922 DA: Prieska, Stayner s.n. sub KG 478/60 (NBG!); Fuller 179 (BOL).
2923 BB: 40 km S Douglas, Plowes 3383.
2924 CA: Hopetown, Fuller 157 (BOL).
2924 DD: Koepal, Bruyns 5117 (BOL).
3121 DD: Kruis v. Bloemfontein, Bruyns 4788b.,
3122 DD: ca. 25 km W Kromrivier, Meve 353*, 355* (NBG!, MSUN).
3123 AA: 18 km N Victoria West, Bruyns 5109.
3123 BD: Richmond townlands, Bruyns 3045 (BOL, NBG).
3123 DD: 13 km E Murraysburg, Bayer 2008* (NBG!); 1 km E Murraysburg, Bayer 2390* (NBG!).
3224 CD: Aberdeen Rd, Pillans 31 (K, BOL, SAM!).
3225 AB: Mountain Zebra Natl. Park, Rooiplat, Meve 457* (MSUN).
3225 BA: 2 mi N Cradock, near Egg Rock, Stayner s.n. sub KG 275/62 (NBG!).
3225 BB: 9 km from Lake Arthur to Spiderkop, Bruyns 1779b (NBG!).

Fig. 35. *Duvalia maculata*. *A* Flower in top view; *B* centre of flower in lateral view; *C* flower in top view; *D* pollinarium; *E* flower in top view; *F* pollination apparatus in top view; *G* pollinarium; *H* flower in lateral view; *I* pollinarium. (*A*, *B* Bruyns 1779b; *C*, *D* Bayer 2008; *E–G* Meve 457; *H*, *I* Meve 353)

8. ***Duvalia modesta*** **N.E. Br.**, in Thiselton-Dyer, Fl. Cap. **4**(1): 1028 (1908).

Type: Eastern Cape, near Aberdeen Road, 24. June 1902, E. Pillans sub Pillans 35 (holo-, K!; iso-, BOL).

Description. Stems 10–40 mm long, 8–15 (−20) mm diam. 4-(5-) angled, ovoid-cylindrical, dark green, glabrous, leaf rudiments ca. 1.5 mm long, stipular glands small, sometimes partially absent. Inflorescence bostrychoid, 1–5 flowers developing successively from around the middle of the stems, emitting excrement odour. Pedicels 8–20 mm long, 1–1.5 mm in diam; sepals 2.5–4 mm long, 1–1.5 mm wide at the base. Corolla 15–25 mm diam., greenish-brown, brown to brownish-purple, lobes 7–9 mm long, 3–6 mm diam., apically replicate, basally less folded or unfolded, basal two thirds of the margins with thin, often bent, purple hairs, 1.5–2.5 mm long, sinuses occasionally covered with these hairs as well, or with papillae, corolla rarely glabrous in total; annulus circular to pentagonal, not or only slightly overtopping the coronal disc, 1–1.5 mm high, 5.5–8 mm diam., uniformly coloured like the lobes, rarely spotted purplish on cream ground, rim of annulus densely covered with hair papillae 50–100 μm long or smooth. Corona ivory, brown, red-brown to dark purple, coronal disc 4–5.5 mm diam., lobes of staminal corona 1–2 mm long, 1–1.5 mm wide in top view. Pollinia yellow to orange-brown, ca. 300 μm long, ca. 200 μm wide, pellucid margin 120–150 μm long; caudicles orange, ca. 100 μm long, translator wings bright orange-brown, hyaline, 200–300 μm long; corpusculum ca. 120 μm long. Follicles 8–13 cm long; seeds ca. 4 mm long, 2.5 mm wide, wing 300–400 μm wide, coma 10–14 mm long. Chromosome number: $2n = 22$ (vouchers: see * in "Specimens examined") (Fig. 36).

Distribution. Eastern Cape; 30°40′–33°45′ S and 24°15′–28°10′ E (Fig. 37).

Habitat. Mainly in Acocks Veld Type 37, False Karroid Broken Veld, but in other veld types also; alt.: 150–1250 m a. s. l.

In comparison to other diploid species, *D. modesta* colonizes a wide range of habitats in a relatively extensive area (Fig. 37). However, karroid habitats are preferred, often in an anthropogenously degraded stage.

Duvalia modesta can be confused with *D. caespitosa*. However, *D. modesta* has delicate stems and flowers with corolla lobes usually not completely replicate to its margins as in *D. caespitosa* (Fig. 36). The small annulus, 1 – 1.5 mm high, is most appropriate to distinguish *D. modesta* from *D. caespitosa* var. *caespitosa*, where the annulus is between 2 and 5 mm in height.

Duvalia pillansii is the most closely related species of *D. modesta*. Indications for the conspecifity of these two taxa, as presumed by Bayer (1984a), could not be found.

Specimens examined

South Africa:

3028 CA: Transkei, Naudes Nek on road to Maclear, Roux 1259* (NBG!).

3221 AD: Aarfontein, Bruyns* 6274 (BOL, MSUN).

Fig. 36. *Duvalia modesta*. *A* Flower in lateral view; *B* flower in lateral view; *C* pollinarium; *D* flower in top view; *E* pollination apparatus in lateral view; *F* pollinarium; *G* budding and flowering stem in top view; *H* pollinarium. (*A* Meve 336; *B*, *C* Meve 334; *D–F* Meve 333; *G*, *H* Meve 329; all specimens from a single locality, "St. Olive", near Graaff Reinet)

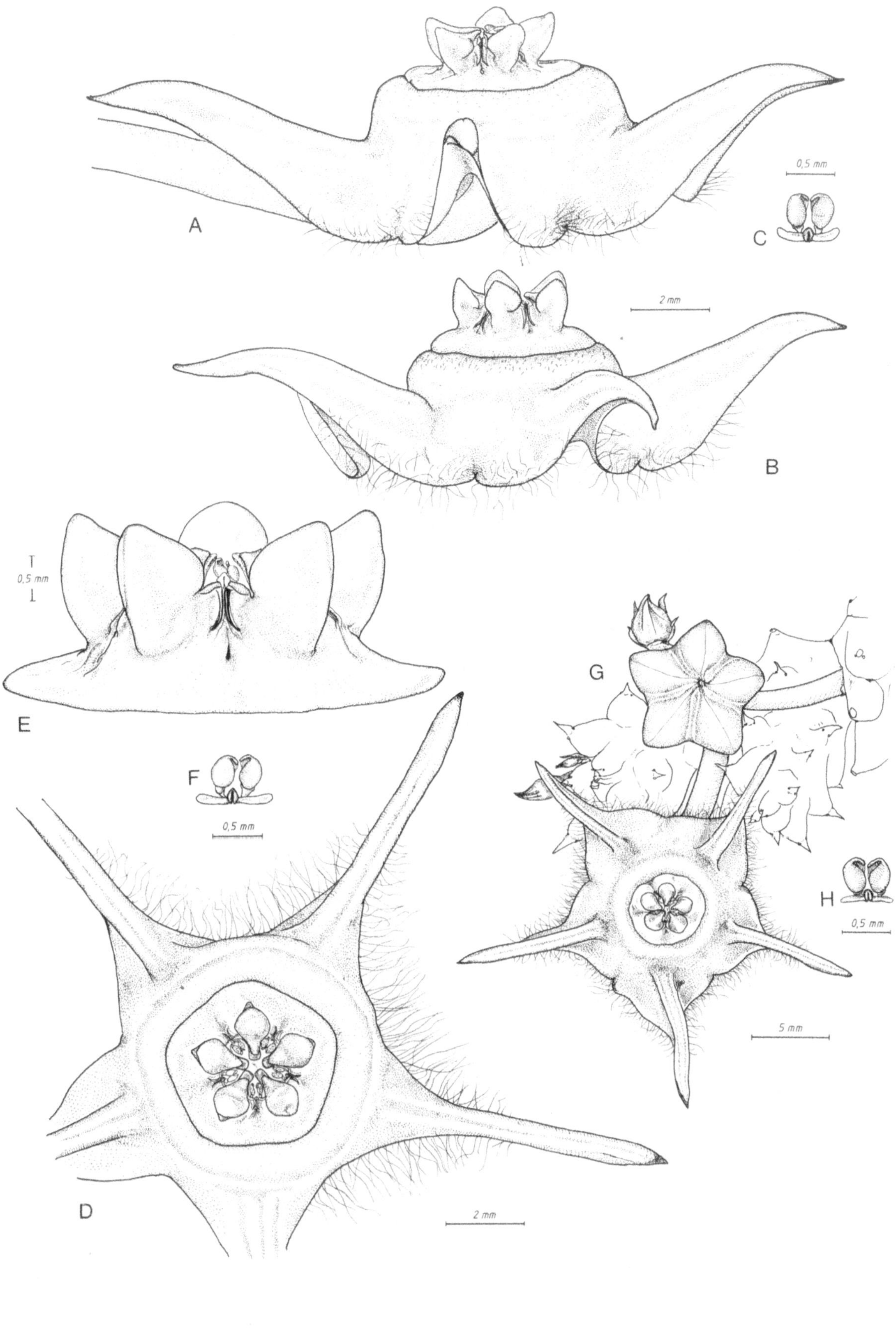
A
B
C
D
E
F
G
H
0,5 mm
2 mm
0,5 mm
0,5 mm
0,5 mm
5 mm
2 mm

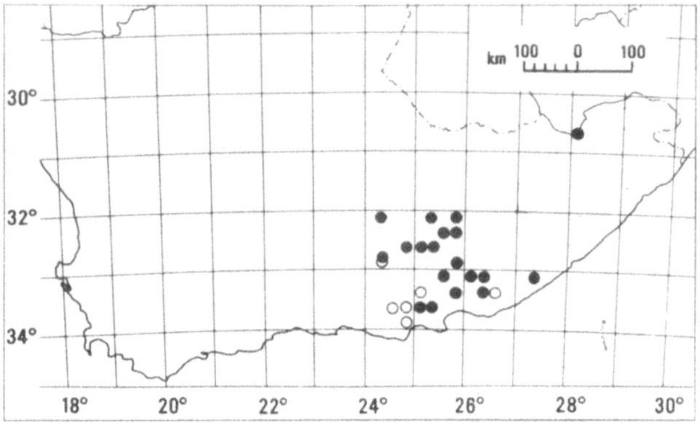

Fig. 37. Distribution of *Duvalia modesta* (●) and *D. pillansii* (○)

3224 AB: NW Graaff Reinet, "St. Olives", BAYER s.n. (NBG); MEVE 328*, 330*, 332*, 333*, 334*, 336*, 337*, 338* (MSUN).
3224 DB: Pearston, Cranemere, BRUYNS 1796a (NBG!).
3225 AB: Cradock, BRUYNS 1581 (NBG!).
3225 BC: Cradock "Mortimer", LOMBAARD s.n. (STEU).
3225 CA: Pearston Rifle-Range, BRUYNS 1588 (NBG!); 5 km N Pearston, MEVE 448*, 449*, 450 (MSUN).
3325 AD: 13 km SW Kirkwood, Steenbokvlakte, BURSEY s.n. sub PLOWES 5235.
3325 BA: Stonefountain, BRUYNS 1566 (NBG!).
3325 BB: Verdun, BRUYNS 1570 (NBG!); 1.5 km E Verdun to Kommadagga, BRUYNS 1594 (NBG!).
3325 BD: Kommadagga, BAYER 634 (NBG!).
3325 CB: Uitenhage, MCEWAN sub NBG 13/57 (NBG!).
3326 AA: Carlisle Bridge, HALL 1607 (NBG!).
3326 AB: N Grahamstown, Dikkop flats, BAYER 630 (NBG!).
3326 AD: 13 km SW Grahamstown, BRUYNS 2168*.

9. ***Duvalia parviflora*** **N.E. BR.**, in THISELTON-DYER, Fl. Cap. **4**(1): 1034 (1908).

Type: Western Cape, Laingsburg. Div., at Witte Poort (3320 BD), 5. May 1906, E. PILLANS sub PILLANS 621 (holo, K!; iso-, BOL).

Description. Stems 10–25 mm long, 10–15 mm diam., spheroid to ovoid, green or bluish-green, glabrous, leaf rudiments minute, < 1 mm diam., stipular glands minute. Inflorescence bostrychoid, 1–6 prostrate to erect flowers developing successively from the basal half of stems, unscented or nearly so. Pedicels 3–8 mm long, ca. 1 mm diam.; sepals 2–3 mm long, 1 mm wide at the base. Corolla 10–15 mm diam., cream, bright yellow to yellowish-green, often with brownish tinged tips of corolla lobes, lobes 3.5–6 mm long, 1.5–2 mm diam., lamina rather convex with incompletely replicate margins, longitudinally furrowed, glabrous; annulus delicate, circular to pentagonal, basally somewhat constricted, 1–2 mm high, 3.5–5 mm diam. Corona ivory to straw- yellow, guide rails occasionally with red tinge, coronal disc raised up to 1 mm above the level of the annulus, 3.5–4 mm diam., rather pentagonal, convex; lobes of staminal corona 0.8–1.4 mm long, 1–1.5 mm in diam in top view, blunt, flat. Pollinia orange-brown, ca. 300 µm long, ca. 200 µm diam., pellucid margin 110–140 µm long; caudicles orange, ca. 100 µm long, translator wings bright orange, hyaline, 300–360 µm long; corpusculum ca. 130 µm long. Follicles 8–12 cm long; seeds 3.5 mm long, 2.5 mm diam., wing ca. 500 µm wide, coma 10–12 mm long. Chromosome number: 2n = 22 (vouchers: see * in "Specimens examined") (Fig. 38).

Distribution. Cape Provinces, Little Karoo, Karoo; 33°20′–33°55′ S and 20°35′–22°30′ E (Fig. 39).

Habitat. Typically found in Acocks Veld Type 25, Succulent Mountain Scrub; alt.: 300–550 m a. s. l.

This rare endemic species of the Little Karoo prefers shrubby, karroid formations of lower altitudes around 400 m. With its small, yellowish flowers and the extremely globose stem morphology, *D. parviflora* is distinctively characterized and appears somewhat isolated taxonomically. It seems to be primitive with regard to floral morphology and flavonoid patterns. Vegetatively, however, this species is advanced with its almost spherical stems and the almost completely reduced leaf rudiments. The total character set of *D. parviflora* is thus of strongly heterobathmic nature. Stem morphology, however, especially in arid areas, obviously is under a much stronger selection pressure by environmental conditions than chemical characters or floral morphology. This idea is confirmed by the common occurrence of similar stems in different *Stapelieae* genera at the same localities. Thus, *D. parviflora* can be identified as a species very close to the ancestral Duvalias.

Specimens examined

South Africa:

3320 DA: E Montagu, HALL 2701 (NBG!).

3321 AD: 8 km E of Ladismith to Calitzdorp, BAYER sub KG 83/77* (MSUN).

3321 AD: Ladismith, JOUBERT s.n. sub KG 290/71* (NBG!, MSUN); 2 km S Ladismith, BAYER 1768 (NBG!); 2 km S Ladismith, JOUBERT s.n. sub KG 49/73 (NBG!).

3321 CA: 3 km S Ladismith to Barrydale, MEVE & KUSCH sub K 1553*.

3321 CB: 11 km E Ladismith, BRUYNS 1408 (NBG!); 11 km S Vanwyksdorp, BRUYNS 4207 (BOL).

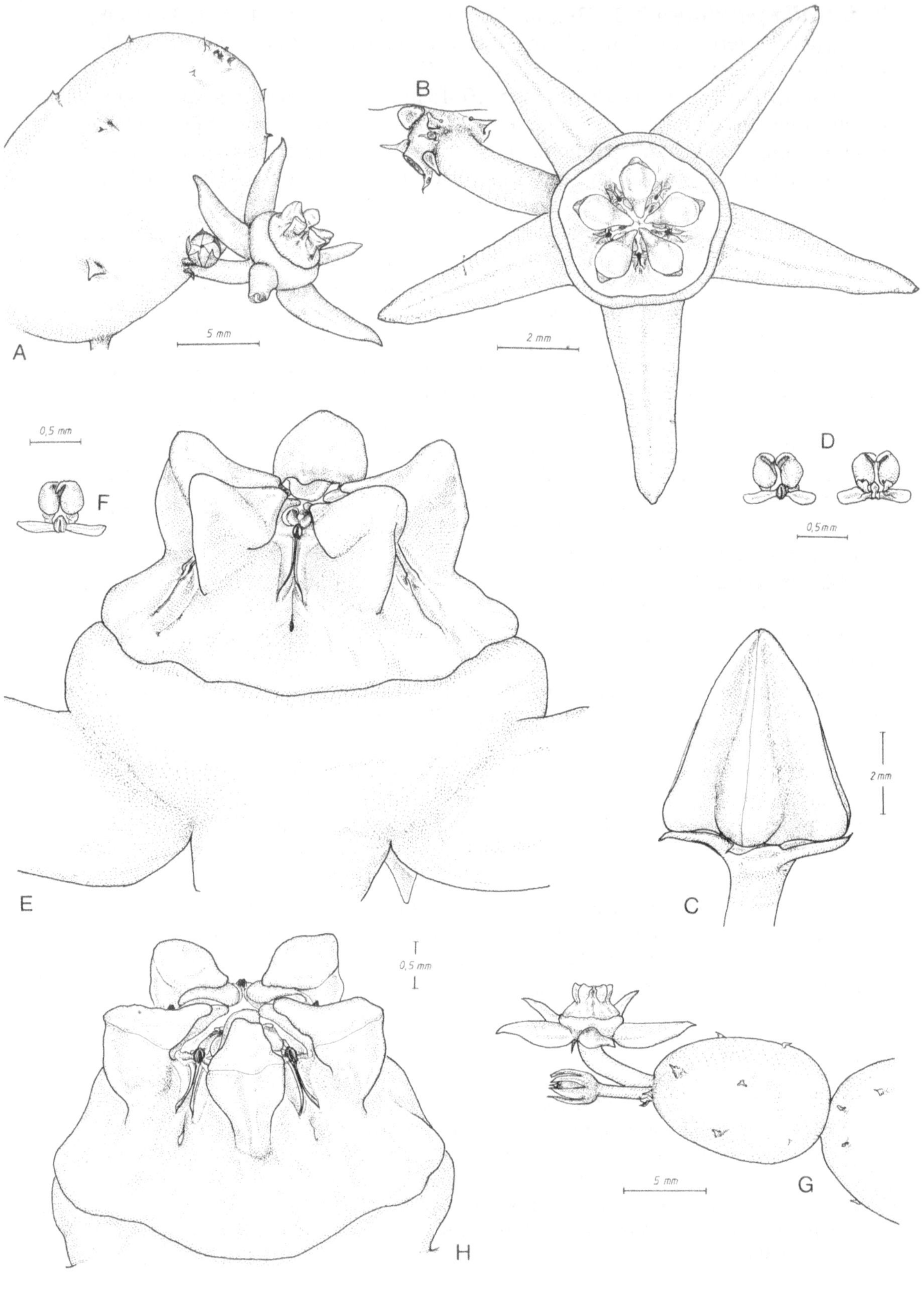
A
5 mm
B
2 mm
C
2 mm
D
0,5 mm
E
F
0,5 mm
G
5 mm
H
0,5 mm

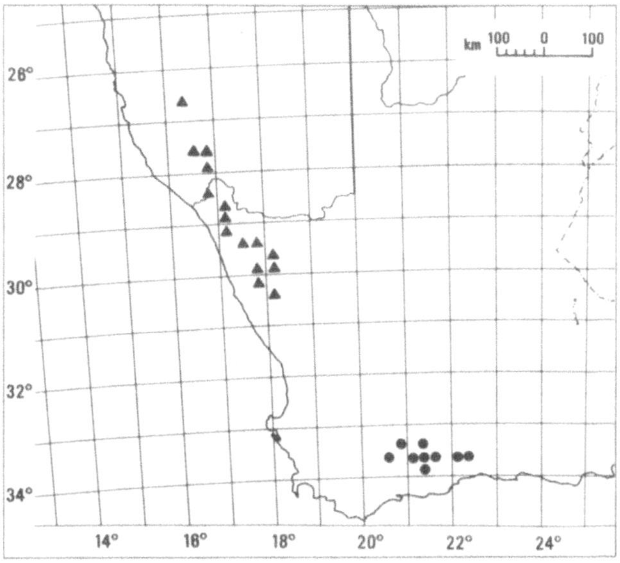

Fig. 39. Distribution of *Duvalia parviflora* (●) and *D. pubescens* (▲)

3321 CD: 4 km S Vanwyksdorp, MEVE & KUSCH sub K 1562* (MSUN); crossing Vanwyksdorp-Riversdale-Barrydale, BRUYNS 4354 (BOL).

3321 DA: 10 km E Vanwyksdorp, BAYER 459 (NBG!); 12 km E Vanwyksdorp, LAVRANOS & PEHLEMANN 20944* (MSUN); 20 km E Vanwyksdorp, LAVRANOS & PEHLEMANN 20947* (MSUN); Assegaaibosch, BRUYNS 4204 (BOL).

3322 CA: Oudtshoorn townlands, LAVRANOS & PEHLEMANN 20603*; 15 km W Oudtshoorn Airport, SCOTT s.n. sub PLOWES 5399 (SRGH).

3322 CB: E Oudtshoorn, Vanwykskraal, BATTEN 553*.

Hybrids seen: *D. parviflora* × *D. caespitosa*

3322CB: 15 km E Oudtshoorn, Vanwykskraal, A. BATTEN sub PLOWES 6281c.

Fig. 38. *Duvalia parviflora*. *A* Flowering stem; *B* flower in top view; *C* bud; *D* pollinaria, left dorsal face, right ventral face; *E* centre of flower in lateral view; *F* pollinarium; *G* flowering stem; *H* pollination apparatus in lateral top view. (*A*, *E*, *F* MEVE & KUSCH sub K 1562; *B*–*D* LAVRANOS & PEHLEMANN 20944; *G*, *H* LAVRANOS & PEHLEMANN 20947 sub K 1507)

10. ***Duvalia pillansii*** **N.E. Br.**, in Thiselton-Dyer, Fl. Cap. **4**(1): 1026 (1908).

Type: Eastern Cape, Aberdeen Div., near Aberdeen road, April 1902, E. Pillans sub Pillans 42 (holo-, K!, iso-, BM!, BOL).

= *D. pillansii* var. *albanica* N.E. Br., in Thiselton-Dyer, Fl. Cap. **4**(1): 1027 (1908). Type: Eastern Cape Province, Albany Div., Vicinity of Grahamstown, Pillans 19 (holo-, K!; iso-, BOL).

Description. Stems 10–30 mm long, 7–14 mm diam., 4-(5-) angled, ovoid to cylindrical, green, glabrous, leaf rudiments ≤ 1 mm long, stipular glands very small, occasionally absent. Inflorescence bostrychoid, 1–4 slightly uplifted flowers developing successively from medium regions of the stems, with slight excrement smell. Pedicels 7–15 mm long, ca. 1 mm diam.; sepals 3–4 mm long, ca. 1 mm wide at the base. Corolla 20–35 mm diam., red to purple-brown, free corolla lobes 7–14 mm long, ca. 8 mm diam., glabrous, lamina basally slightly convex, tapering into tips with more or less completely replicate margins, margins with simple, often bent hairs, 2–3 mm long, purplish; annulus circular to slightly pentagonal, peripherally subsequentally merging into the corolla lobes, 7–10 mm diam., 2–2.5 mm high, white to creamish, occasionally the inner rim only, covered with hair papillae, 50–100 µm long. Corona cream to straw- yellow, coronal disc ± circular, ca. 5 mm diam., lobes of staminal corona ca. 1–1.5 mm long, ca 1.5 mm wide in top view. Polliniam orange, ca. 350 µm long, 240 µm diam., pellucid margin ca. 180 µm long; caudicles orange, ca. 100 µm long, translator wings bright orange, hyaline, ca. 300 µm long; corpusculum ca. 160 µm long. Follicles ca. 8 cm long; seeds 4.5–5.5 mm long, ca. 3 mm diam., wing often emarginate, 0.5–1 mm diam., coma 10–12 mm long. Chromosome number: 2n = 22 (vouchers: see * in "Specimens examined"). (Fig. 40).

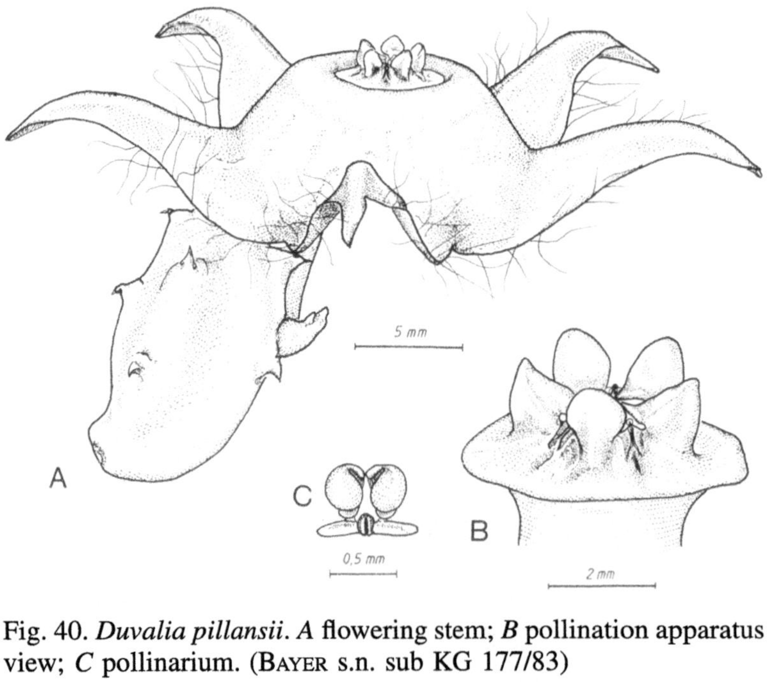

Fig. 40. *Duvalia pillansii*. *A* flowering stem; *B* pollination apparatus on its column in lateral view; *C* pollinarium. (Bayer s.n. sub KG 177/83)

Distribution. Eastern Cape; 32°50′–33°55′ S and 24°15′–26°40′ E (Fig. 37).

Habitat: Mainly in the Acocks Veld Type 23, Valley Bushveld; alt.: 150–750 m a. s. l.

Duvalia pillansii probably represents the least well understood species of the Cape. Its distribution area (Fig. 37) overlaps with the one of its sister species, *D. modesta*. However, no signs of interbreeding have been found. *Duvalia pillansii* can be easily distinguished from *D. modesta* by its wide corolla lobes and its unusual bright yellow corona.

Specimens examined

South Africa:

3324 DA: Baviaanspoort, STAYNER s.n. sub KG 200/71 (NBG!).

3324 DD: 0.25mi E Hankey north of school, BAYER 677, 760* (MSUN; NBG!); 4 km E Hankey, BAYER s.n. sub KG 177/83*, NBG; 1 km NE Hankey, BRANDHAM & CUTLER s.n. sub Kew 35376 (K!).

3325 AC: Sapkamma, s. coll. sub KG 204/60 (NBG!).

11. ***Duvalia polita*** **N.E. Br.**, Gard. Chron., ser. nov. **6**: 130 (1876).
Type: Cultivar Royal Botanic Gardens Kew (holo-, K!).
= *D. polita* var. *polita* f. *intermedia* A.C. White & B. Sloane, Stap. **2**: 755 et Stap. **3**: 1144 (1937). Type: Icon. (Photo) White & Sloane, Stap. **2**: 756, fig. 751, lower flower (lecto-, designated here).
= *D. polita* var. *transvaalensis* (Schltr.) A.C. White & B. Sloane, Stap. **2**: 754 (1937). Basionym: *D. transvaalensis* Schltr., Bot. Jahrb. Syst. **20** (Beibl. 51): 54 (1895). Type: Transvaal, sandige Stellen bei Klipdam, Schlechter 4498 (holo-, BOL).
= *D. transvaalensis* var. *parviflora* L. Bolus, Ann. Bol. Herb. **1**: 194 (1915). Type: "Seringa" near Naboomspruit '3750', Galpin 8467 (holo-, BOL).
= *D. polita* var. *parviflora* (L. Bolus) A.C. White & B. Sloane (*comb. illeg.*), Cact. Succ. J. (US) **14**: 159 (1942). Type as for preceeding taxon.
= *D. dentata* N.E. Br., Bull. Misc. Inform. (Kew) 1895: 265 (1895). Type: Bechuanaland (Botswana), 30 miles NW of Koobie, 1863, Baines s.n. (holo-, K!).

Description. Stems above ground 20–100 mm long, 7–15 mm diam., 6-angled, cylindrical, tapering towards the apex, green, glabrous, leaf rudiments narrow, acute, 4–6 mm long, stipular glands elongated, ca. 0.5 mm long; subterranean stems (stolons) 10–100 mm long, cylindrical, pencil-thick, 5–8 mm diam., ivory (the horizontally growing, subterranean parts can constitute more than 50 % of the plant's biomass). Inflorescence dichasial or bostrychoid, 1–4 prostrate to erect flowers developing successively from basal regions of the stems, with a strong excrement stench. Pedicels 15–25 mm long, ca. 1 mm diam.; sepals 4–5 mm long, 1–2 mm wide at the base. Corolla 20–35 mm diam., lobes 10–15 mm long, 7–10 mm wide, lamina slightly convex with tips raised above the level of the annulus, reddish-brown to purple, blackish-brown, or spotted with markings in the above mentionend colours on yellowish basal coloration, glabrous, brilliant, margins with vibratile, purplish clavate hairs, 0.5–2.5 mm long, basally or up to 2/3 of their lengths, (these hairs rarely absent); annulus circular, rarely slightly pentagonal, 2.5–5 mm high, 8–12 mm diam., spotted or unspotted, upper rim with hair papillae, 50–200 µm long, peripherally overflowing into the basally fused corolla lobes. Coronal disc circular to pentagonal, 4.5–6.5 mm diam., reddish-brown, lobes of the staminal corona 1.5–2.5 mm high, ca. 2 mm long., bright reddish-brown, outer regions usually flattened, stout, spoon-shaped in top view. Pollinia yellow to reddish-brown, ca. 400 µm long, 270 µm diam., pellucid margin ca. 200 µm long; caudicles orange, ca. 120 µm long, translator wings bright yellow to bright orange, ca. 350–400 µm long; corpusculum ca. 200 µm long. Follicles up to 20 cm long; seeds 5–7.5 mm long, ca. 4–5 mm diam., wing strongly developed, 1–1.3 mm diam., coma 15–20 mm long. Chromosome number: heteroploid with $2n = 22$, $2n = 44$ (vouchers: see * in "Specimens examined") (Figs. 41, 42).

Distribution. Angola, Botswana, Malawi, Mozambique, Namibia, South Africa, Swaziland, Zambia, Zimbabwe; ca. 16°00′–29°15′ S and 15°15′–35°00′ E (Fig. 43).

Habitat. Often in Acocks Veld Type 16, Kalahari Thornveld, or in Mopane-(*Colophospermum*-)bush, or -woodland, usually associated with slightly brackish soil; alt.: 1000–1600 m a. s. l.

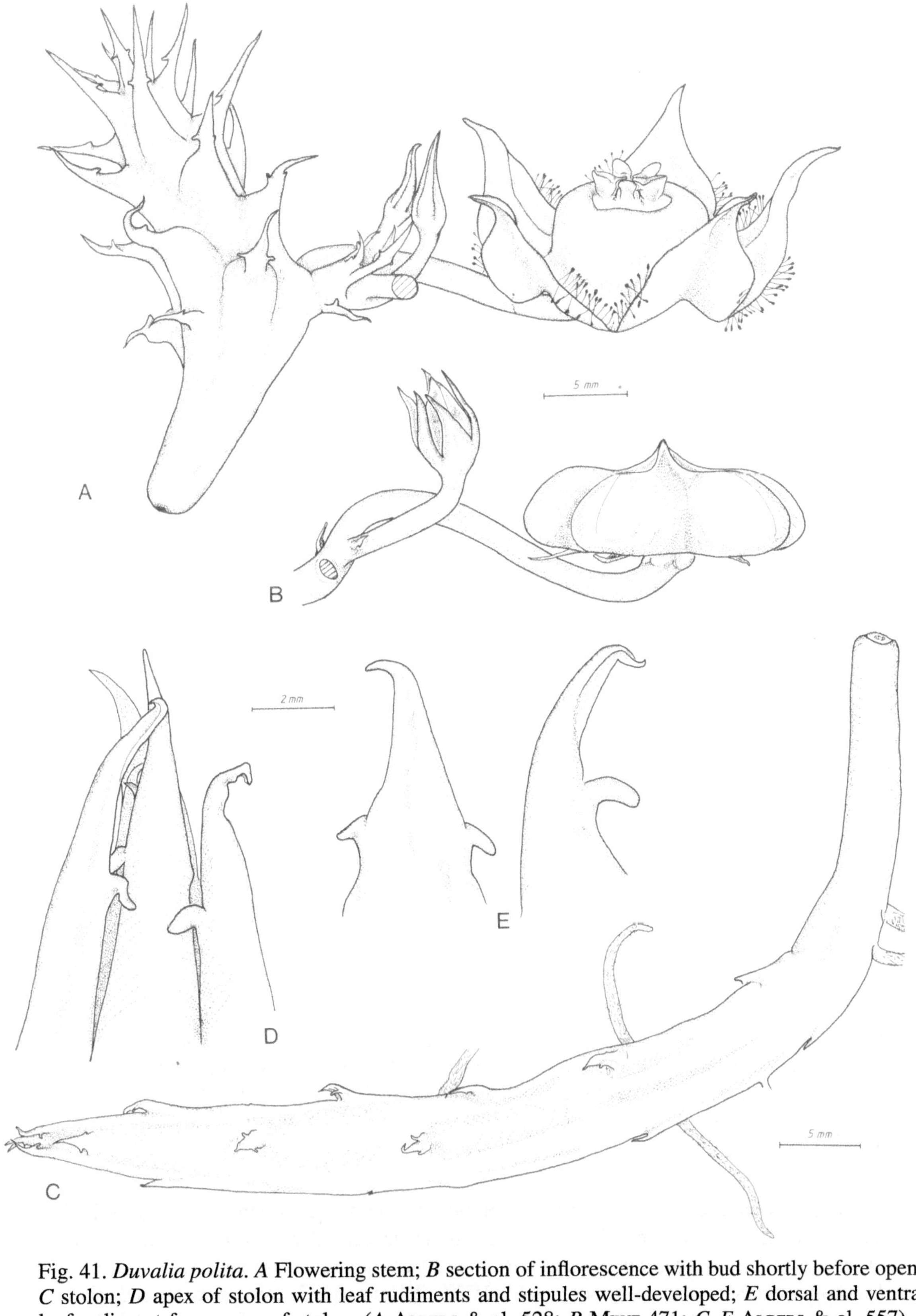

Fig. 41. *Duvalia polita*. *A* Flowering stem; *B* section of inflorescence with bud shortly before opening; *C* stolon; *D* apex of stolon with leaf rudiments and stipules well-developed; *E* dorsal and ventral of leaf rudiment from apex of stolon. (*A* Albers & al. 528; *B* Meve 471; *C–E* Albers & al. 557)

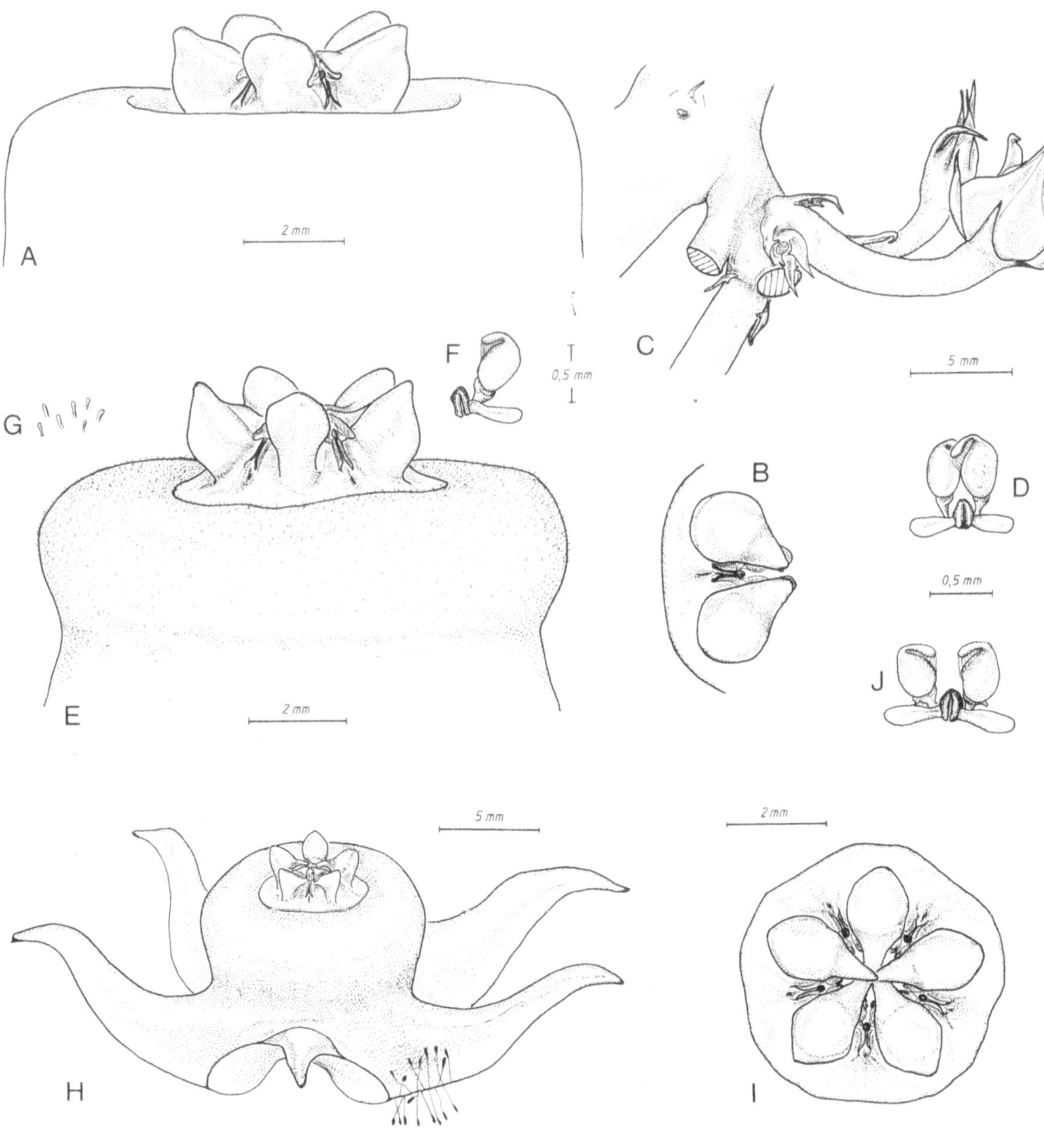

Fig. 42. *Duvalia polita*. *A* Rim of annulus and corona in lateral view; *B* section of corona in top view; *C* budding inflorescence; *D* pollinarium; *E* centre of flower in lateral view; *F* pollinarium; *G* hair papillae from upper rim of annulus; *H* flower; *I* pollination apparatus; *J* pollinarium. (*A–D* Albers & al. 557; *E–G* Gerbaulet & Struck 23632; *H–J* Albers & al. 565)

The systematics of this variable species has confused even the taxonomists dealing with this complex: Schlechter (1885) described glabrous individuals of *D. polita* as *D. transvaalensis* and L. Bolus, 20 years later, followed with a variety, *D. transvaalensis* var. *parviflora*. In 1937, White & Sloane recognized that *D. transvaalensis* does not represent a good species and therefore reduced it to

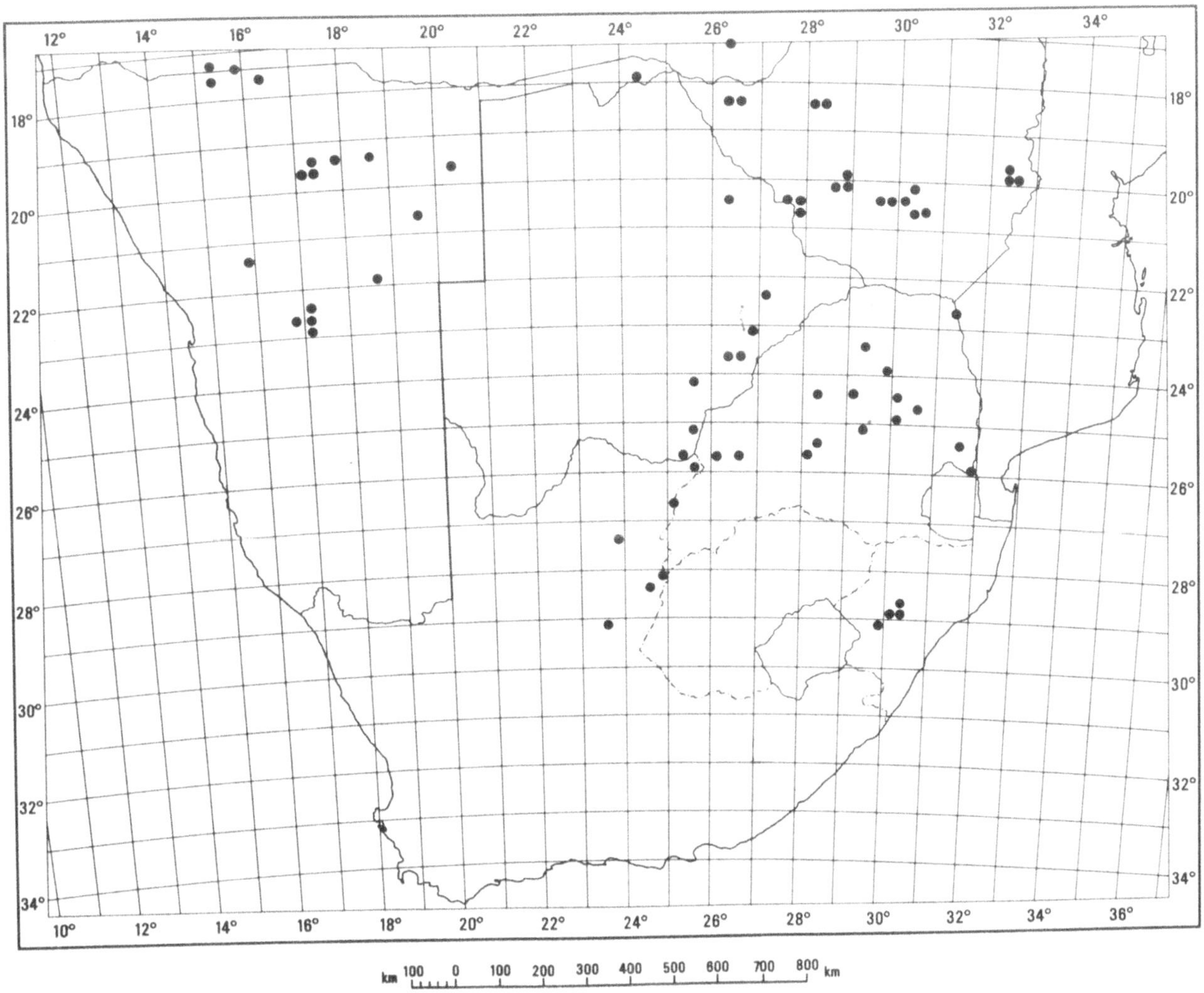

Fig. 43. Distribution of *Duvalia polita* south of 17° latitude

the rank of a variety, combining correctly *D. polita* var. *transvaalensis*. However, in 1942, White & Sloane erroneously supposed that they acted wrongly and in contradiction to the Code while publishing this combination, because the Code dictates that the name of an infraspecific taxon has to be preserved when it is transferred to another species (ICBN 1935, Art. 55). White & Sloane (1942) misinterpreted this article, and published the "corrected" combination, *D. polita* var. *parviflora*. However, *D. polita* var. *parviflora* is illegitimate, because of the autonym var. *transvaalensis*. White & Sloane overlooked that this was automatically established with the new variety of Bolus, *D. transvaalensis* var. *parviflora*, and has priority against var. *parviflora* [ICBN (1995), Art. 11.6].

Open, shrubby or woodland formations of *Acacia* or Mopane can be described as the most characteristic habitats of this species widely distributed in the subtropical summer rainfall areas of southern Africa. In *Acacia*-communities *D. polita* competes mainly with grasses, which are less dominant in the

Colophospermum localities. However, while the latter are more open and under less competition, they usually have the disadvantage of brackish soil. The production of a considerable mass of subterranean stems (stolons/rhizomes) by *D. polita* plants can be understood as an adaptation to bush fires.

Apart from this specialized growth form, *D. polita* exhibits primitive characters with elongated, 6-angled and "leafy" stems (Fig. 41A), and, in its floral morphology, the shape of the corolla lobes and the clavate papillae on the adaxial surface of the corolla lobes (Fig. 42G) as well as the very long vibratile, clavate hairs (up to 2.5 mm long!) along the margins. Also, the wide wings of the seeds are unusal for the genus (Fig. 12A, B), pointing to a close relationship with *Huernia*, where such seeds are widely distributed (Meve, unpubl.). Despite these morphological deviations, *D. polita* is undoubtly a true member of sect. *Duvalia*. However, it occupies a somewhat isolated position between the two sections, a placement supported by its far-reaching crossing incompatibilities with other *Duvalia* species. Despite its intermediate distribution area, *D. polita* thus cannot be considered as a link between the two sections.

No forms or varieties of significant geographical distribution are worth recognizing. There is only a tendency that plants possessing clavate hairs on the flowers are more frequent in the eastern part of the distribution area than in the western one. In Namibia, *D. polita* usually displays a blackish-brown corolla without maculation. In the Transvaal and in Zimbabwe individuals with maculated annuli dominate. However, maculated and unmaculated flowers have been observed within the same populations (Zimbabwe: Bulawayo). The recognition of infraspecific taxa within this species therefore is no longer justified.

The occurrence of isolated diploid cytotypes in the Transvaal and in Zimbabwe remains without phenotypical effects.

Specimens examined

Angola:
Rocadas, Leach & Cannell 4043 (NBG).

Botswana:
2026 AD: 100 km W Serowe, N Lephephe, Snyman & Noailles s.n. (PRE!).
2326 BB: Mahalapye, Bayliss 1858 (NBG!).
2326 CB: Debeeti, Plowes 4384 (SRGH!).
2425 BA: 5 km E Dodi (N of Gabarone), Plowes 6702 (SRGH!).
2525 CB: Lobatse, Thornpark, Bayer 763 (NBG!).

Namibia:
1715 BD: Ovamboland, Oshikango, Rodin 8922 (SRGH!).
1715 CB: 25 km from Oshakati towards Ogongo, Van Jaarsveld 2825 (NBG!).
1715 DD: 30 km SE Oshakati, Bruyns 4106 (BOL).
1716 : Ovambo, Oininive, Giess 12432 (WIND!).
1724 CD: 100 m from edge of Lake Liambezi, Bruyns 2294 (WIND!); N Lake Liambezi, Zilitenfe, Bruyns 2333 (WIND!).
1917 AD: Hereroland, Otavi, Dinter 1382 (K!).
1917 BD: Nosib Tsu 655, Bruyns 2340 (WIND!).
1917 CA: Farm Kumkauas (Skoll), Giess 12590 (WIND!).
1917 CB: Otavi, Dinter 1660 (SAM!).

1918 BC: Osmoor, BRUYNS 4138 (BOL).
1920 DA: Bushmanland, Tsumkwe, GIESS s.n. sub PLOWES 4386 (SRGH!).
2115 BD: Omaruru, OTZEN s.n. sub 579/38 (NBG!).
2118 DC: Steinhausen, Farm Wunderland, ALBERS & GRABOW sub K 1605*; Namibia, Steinhausen, near Witvlei, Farm Wunderland, BRUYNS 2275 (WIND!); Namibia, Steinhausen, near Witvlei, Farm Wunderland, LEUENBERGER & al. 3121 (WIND!, B!).
2119: southern Bushmanland, HINES 651* (WIND!).
2217 CC: Auasgebirge, Lichtenstein, DINTER 4464 (B!); Windhoek, Farm Keres, VENTER s.n. (WIND!).
2216 DB: near Windhoek, Farm Friedenau, SANDERS s.n. (WIND!).

South Africa:
2329 AD: 32 km N Pietersburg, PLOWES 2635 (SRGH!).
2329 DD: Transvaal (NE), Haenertsburg, Boune, BALKWILL s.n. sub K 1574*.
2428 AD: between Warmbad and Potgietersrus, ALBERS s.n. sub K 790*.
2430 AC: 25 km N Burgersfort, LANCESTER 195 (SRGH!).
2430 CA: Steelpoort, HENDERSON s.n.* (MSUN).
2430 DA: 2 km from Ohrigstad, Burgerfont, BRUYNS s.n. sub KG 105/76 (NBG!).
2525 BA: Ootse Village (outside Moeding College), WOOLARD 1362 (SRGH!)
2525 DC: 12 km W Mafeking, BRUYNS 2013 (PRE!), Mafeking, BAYER s.n. sub KG 71/71 (NBG!).
2526 CA: Zeerust, LEENDERTZ 4353 (K!).
2526 DA: Zeerust-Rustenberg, 1 km W junction Rickertsdam/ Swarterruggans, GERBAULET & STRUCK 23632* (MSUN).
2528 AD: Pretoria distr., Pienaars River, KRANSDORFF s.n. sub NBG 588/48 (NBG!); 1 km E Hammannskraal, MEVE 469, 470*, 471*, 474* (MSUN).
2528 CA: Pretoria, KRANSDORFF sub NBG 588/72 (NBG!).
2529 AB: Transvaal, Groblersdal, Bushveld, MORRIS s.n. sub NBG s.n. (NBG!).
2531 CB: Krüger Nat. Park, Malelane, CODD 5512 (K!, SRGH!).
2625 A : "Bophutatswana", BRUYNS s.n. (NBG!).
2723 BD: Kuruman, HALL 1724 (NBG!).
2824 BB: 5 km NW Warrenton, FRANDSEN s.n. in DP 4210 (SRGH!).
2830 CB: Zululand, Tugela Ferry, PLOWES 3916 (SRGH!).
2830 CC: Weenen, Bushmans River, GREEN s.n. sub PLOWES 3914 (SRGH!).
2830 CD: 5 mi SE Muden, CODD 8599a (K!).
2923 BA: Florisfontein, BRUYNS 3071 (BOL).
2929 BB: Estcourt, SIDEY 186 (MO!).

Swaziland:
2531 DD: Bordergate - Komatipoort Rd., LEACH & BAYLISS 10605 (PRE!).

Zambia:
1726 AB: Kalomo, FANSHAWE 9229 (SRGH!).

Zimbabwe:
1826 AD: ca. 2 km SW Hwange, ALBERS, LIEDE & MEVE 527.
1828 BC: 1 km S Sengwa, Lutope River junction Gokwe, GUY s.n. sub PLOWES 5027 (SRGH!).
1928 DD: 18 mi NW Bulawayo, LEACH 1386 (SRGH!).
1932 CD: Birchenough Bridge, HALL s.n. sub NBG 375/57 (NBG!).

1932 DC: 16 km N Birchenough Bridge, Plowes 3014 (SRGH!).
2027 BC: 24 km NW Plumtree, Brewer s.n. sub Plowes 4008 (SRGH!).
2027 BD: 2 km N Plumtree, Meve, Erwee & Liede 544* (MSUN); SW Zimbabwe, Plumtree (suburb), Meve, Erwee & Liede 557* (MSUN).
2028 BA: Bulawayo, Hyde Park, Albers 2564*; Albers, Liede & Meve 528*, 530*, 531*, 532*, 533* (MSUN).
2028 BB: Heanys Junction, Albers 558* (MSUN).
2029 BC: 140 km W Masvingo, Meve, Erwee & Liede 563*.
2029 BD: 95 km W Masvingo, Albers & Meve 565* (MSUN).
2030 AB: near Mashaba (Nyanagedzi River), Leach 5748 (SRGH!).
2030 DB: 70 km S Masvingo, Plowes 3915 (SRGH!).

Localities imprecise or not found: Botswana, NE Mopipi, Peterhouse 579 (SRGH!); Okavango swamps, Smith 218 (SRGH!). Namibia, Damaraland, Capt. Een s.n. (K!); Musundra, Hardy 5657*; NE of Karakuwise, Cigarette, Maguire 2329 (NBG!); 2217 CA:Lake Liambezi near Windhoek, Müller 2/83* (WIND!). South Africa, Transvaal, Sekuniland, Hall s.n. sub NBG 530/56 (NBG!); 20 mi W Limhangeni (N'tamba Valley), Houting s.n. NBG 43/44 (NBG!). Zimbabwe, Sebi Valley, Hot Springs, Plowes 2474*; S Gwelo, Somabula flats, Plowes 2484 (SRGH!).

12. ***Duvalia pubescens*** **N.E. Br.**, in Thiselton-Dyer, Fl. Cap. **4**(1): 1029 (1908).

Type: Northern Cape, Little Namaqualand, Ayres s.n. sub Pillans 94 (holo-, K!; iso-, BOL).

= *D. pubescens* var. *major* N.E. Br., in Thiselton-Dyer, Fl. Cap. **4**(1): 1028 (1908). Type: Northern Cape, Little Namaqualand, Scully s.n. (holo-, K!).

= *D. elegans* var. *namaquana* N.E. Br., in Thiselton-Dyer, Fl. Cap. **4**(1): 1028 (1908). Type: Northern Cape, Little Namaqualand, March 1875, Barkly 34 (holo-, K!).

Description. Stems 10–50 mm long (occasionally longer in cultivation), 7–22 mm wide, 4–5-angled, ovoid to cylindrical, dark green to grey-green, glabrous, leaf rudiments 1–2 mm long, stipular glands well developed. Inflorescence bostrychoid, 1 to several flowers developing successively from near the stem base, with mild to intense carrion smell. Pedicels 10–25 mm long, ca. 1 mm diam.; sepals 3–4 mm long, ca. 1.5 mm wide at the base. Corolla 18–32 mm diam., lobes 7–13 mm long, 2.5–3.5 mm high at the base of the widely replicate lamina, bright reddish-brown to purplish, adaxially pubescent, sometimes only on the basal half, hairs 0.5–1.4 mm long, erect, whitish to purplish; annulus circular to pentagonal, apically flattened, 6.5–12 mm diam., 1.5–3 mm high, pubescent, hairs 0.3–1 mm long, whitish to purplish, simple, marginally often maculated. Corona bright orange-brown, reddish-brown (in Namibia also white), coronal disc 4.5–6 mm in diam, circular to pentagonal, staminal corona lobes 1–2.5 mm high, 1.2–2 mm long, dorsal parts horizontally flattened to vertically erect, obtuse to acute. Pollinia 350–450 µm long, 250–300 µm wide, reddish-brown, germination mouth ca. 140 µm long; caudicles 150–200 µm long, orange, translator wings bright orange, 350–450 µm long; corpusculum ca. 200 µm long. Follicles 8–14 cm long; seeds 4.5–5.5 mm long, ca. 3 mm wide, wing ca. 500 µm wide, coma 10–15 mm long. Chromosome number: $2n = 44$ (vouchers: see * in "Specimens examined") (Fig. 44).

Distribution. Northern Cape (Namaqualand, RSA), southern Namibia; 26°35′–30°25′ S and 16°20′–18°15′ E (Fig. 39).

Habitat. Typically found in the Acocks Veld Type 33, Namaqualand Broken Veld, alt.: 650–1000 (1350) m a. s. l.

Duvalia pubescens is endemic to the semi-arid regions north of the Olifantsrivier, which receives winter rainfall only (Fig. 39). Here, it is a character species of the "Namaqualand-Namib-Domain" of the "Succulent Karoo Region" (Jürgens 1991). *D. pubescens* prefers sandy to sandy-loamy substrates. In Namibia (Aus) it was also found growing in crevices on granite (I. Pehlemann-Brase, pers. comm.). Apart from this ecological behaviour, it shows distinctive chemical (kaempferol present in *D. pubescens,* lacking in *D. caespitosa*, Table 6) and morphological characters (pubescent corolla surface). *Duvalia pubescens* and *D. caespitosa* are strictly allopatric and every flowering individual can easily be assigned to the one or the other taxon. Therefore, this species cannot be merged with *D. caespitosa* as suggested by Bayer (1977, 1984a) and Bruyns (1981a). However, *D. pubescens* undoubtly represents the sister species of *D. caespitosa* (both with $2n = 44$ and similar flower morphology).

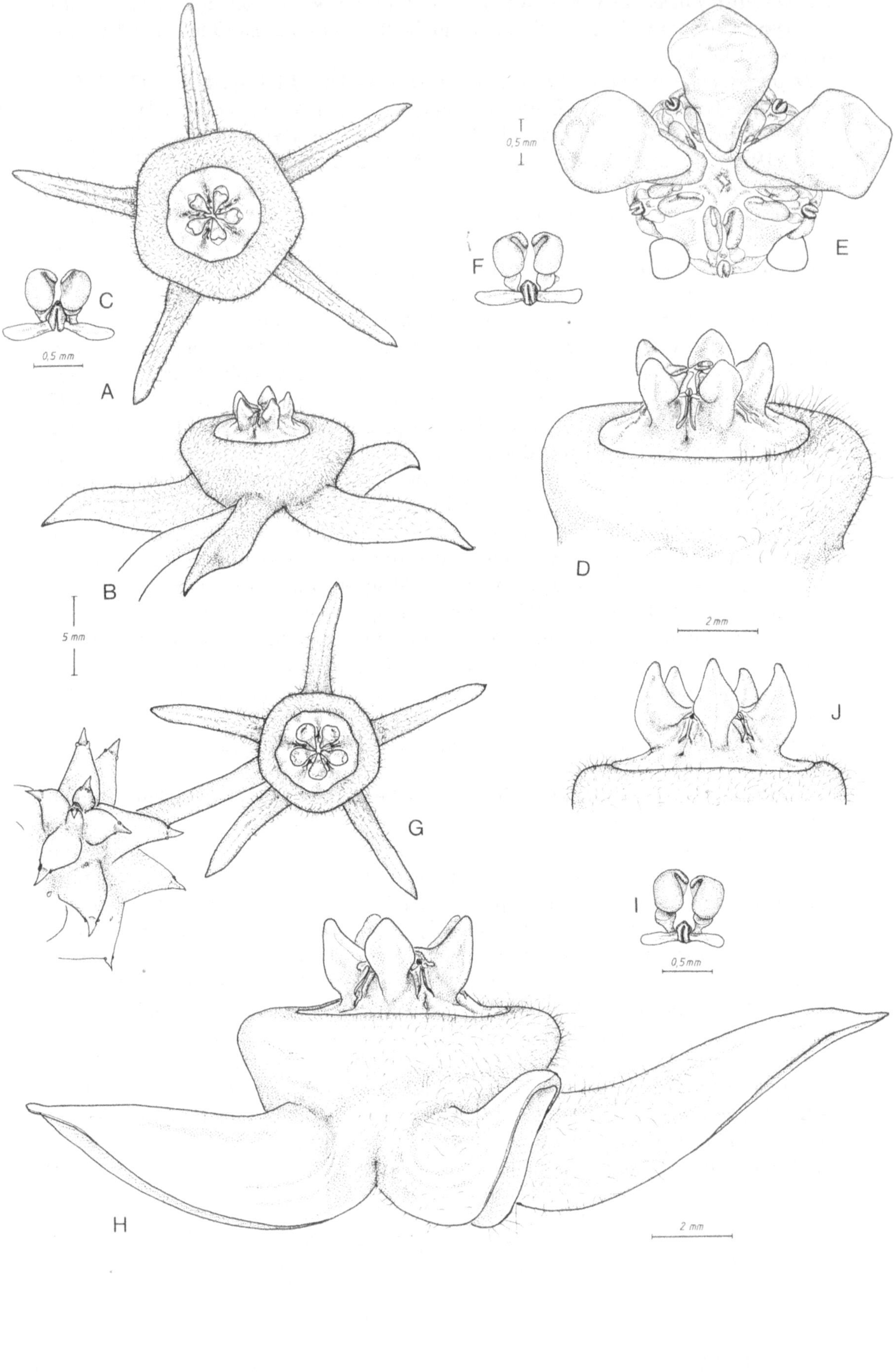
A
B
C
D
E
F
G
H
I
J
0,5 mm
0,5 mm
0,5 mm
5 mm
2 mm
2 mm

Specimens examined

Namibia:

2616 CB: Aus, Dinter 6218 (B);
2716 DA: Sebrafontein (Namib side), Bruyns 3939 (BOL).
2716 DB: Sebrafontein, Bruyns 3909 (BOL).
2716 DD: Rosh Pinah, Namuskluft, Lavranos & Pehlemann 19944* et 20772* (MSUN).
2816 BB: Kahanstal, Dinter 8112 (B!).

South Africa:

2816 BD: Richtersveld, top of Hellskloof, Albers, Kusch & Meve sub K 1355* (MSUN); Hellskloof, Albers & Meve 33* (MSUN); Richtersveld, 1 km SE of top of Hellskloof, Bayer 1506a (NBG!); Richtersveld, Modderfontein near Kuboos, Compton s.n. sub NBG 1070/48 (NBG!); Hellskloof, Jürgens 22534*.
2817 CA: Richtersveld, Summit Cornellsberg, Bruyns 3226*.
2917 AA: Richtersveld, 10 km E Lekkersing, Plowes 5418 (SRGH).
2917 BD: 8.5 km N Concordia, Meve 218*, 219*, 221*, 222*, 224, 225, 226, 227, 228 (MSUN); 5.5 km N Concordia, Meve 230 (MSUN); 5.1 km N Concordia, Meve 232* (MSUN).
2917 CB: 4.8 km N Concordia, Meve 165*, 166*, 168* (MSUN).
2917 DD: N Kamieskroon, Buffels Rivier, Albers, Kusch & Meve sub K 1374* (MSUN).
2918 CA: 18 mi SE Springbok to Gamoep, Wisura 2922 (NBG!); 28 km E Springbok, Bayer s.n. sub Plowes 4585 (SRGH).
2918 CC: 25 km SE Springbok, Bayer 741 (NBG!).
3017 BB: 1 m N Kamieskroon, Leistner 750 (PRE!); 3 km E Kamieskroon, Meve 152, 153*, 154* 154* (MSUN).
3018 AC: E Garies, Studers Pass, Bayer 635 (NBG!).
3118 DC: (?) Nardouw, s. coll. s.n. sub NBG 572/41 (NBG!).

Locality imprecise: South Africa, N Cape, between Springbok and Kamieskroon, Herre s.n. (NBG!).

Fig. 44. *Duvalia pubescens*. *A* Flower in top view; *B* flower in lateral view; *C* pollinarium; *D* centre of flower; *E* pollination apparatus in top view (two staminal corona lobes removed); *F* pollinarium; *G* flowering stem in top view; *H* flower in lateral view; *I* pollinarium; *J* centre of flower in lateral view. (*A–C* Meve 224; *D–F* Albers & al. sub K 1372); *G–I* Lavaranos & Pehlemann 19944 sub K 1680; *J* Albers & Meve 33)

13. ***Duvalia vestita*** Meve, Kakteen Sukk. **39**: 194 (1988).

Type: Western Cape, SW of Bredasdorp, 2 km N Wiesdrift, 26. Nov 1986, Meve & Liede 397 (holo-, K; iso-, MSUN).

Description. Stems 15–70 (–100) mm long, 8–18 mm wide, 4-angled, ± cylindrical, plagiotropous to nearly orthotropous, usually rhizomatous, dark green, glabrous, leaf rudiments ≤1 mm long, stipular glands minute or absent. Inflorescence bostrychoid, 1–4 prostrate to slightly erect flowers developing successively from stem base, with intense excrement and carrion stench. Pedicels 10–20 mm long, ± 1 mm diam.; sepals 2–3 mm long, basally ± 1 mm wide. Corolla 20–30 mm diam., reddish-brown to blackish-purple, ± brillant, lobes 7–13 mm long, 2.5–4 mm high at the base of the widely replicate lamina, adaxially pubescent to pilose on basal half to whole lamina, hairs up to 2.2 mm long, purplish; annulus ± pentagonal, apically flattened, 7–10 mm diam., 1–3 mm high, hairy, hairs 1.2–2 mm long, purplish. Coronal disc ca. 6–7 mm diam., penta- to decagonal, reddish-brown to purple, half covering the inner rim of annulus, staminal corona lobes 1.5–2 mm high, 1.6–2.2 mm long, ± orange-brown. Pollinia yellow- to brownish-orange, ≤ 400 µm long, ca. 270 µm wide, pellucid margin ca. 180 µm long; caudicles 140–200 µm long, orange, translator wings 350–450 µm long; corpusculum ca. 200 µm long. Follicles up to 16 cm long; seeds ca. 5–6 mm long, 4–4.5 mm wide, wing ca. 1 mm wide, coma 15–18 mm long. Chromosome number: 2n = 44 (vouchers: see * in "Specimens examined") (Fig. 45).

Distribution. Western Cape, Bredasdorp to Mosselbaai (33°50′–34°45′ S and 19°20′–24°50′ E (Fig. 31).

Habitat. Lowland species, predominantly in Acocks Veld Type 47, Coastal Fynbos; alt.: 50–250 m a.s.l.

Duvalia vestita has been confused with *D. pubescens*, even though it is closest to *D. elegans* geographically as well as morphologically (cf. Meve 1988). The possibility that the species represents a hybrid with, e.g., *D. caespitosa* and *D. elegans* as parental species, cannot be ruled out. However, *D. vestita* today deserves the status of a good, stable species, because reproducing populations from several different localities are known. The stable karyological situation observed within these populations underlines this status. If *D. vestita* was a cross between a diploid (*D. elegans*) with a tetraploid (*D. caespitosa*) species, a triploid genome should be expected. The tetraploidy of *D. vestita* presumably originates from an unreduced gamete (of *D. elegans*?), with $n = 22$, fused with a normal gamete of *D. caespitosa* to produce a tetraploid zygote, and initiating the first *D. vestita* seed from which the first plant arose. To ensure reproduction of this new genome, self-fertility is the most urgent presupposition, and, indeed, *D. vestita* proved self-fertile in my experiments (Table. 7). Typically, such hybrids, as well as newly formed autopolyploids (cf. Ehrendorfer 1980), are unable to compete with their parents in the same habitat. The distribution areas of the three species concerned show that *D. caespitosa* is peripatrically distributed with *D. vestita,* while *D. elegans* and *D.*

Fig. 45. *Duvalia vestita*. *A* Flower in longitudinal section; *B* bud; *C* flower in top view; *D* pollination apparatus; *E* pollinarium; *F* section of flower in top view; *G* pollinarium; *H* centre of flower; *I* pollinarium; *J* stem in lateral view; *K* flower in lateral view. (*A–E* Hartmann 12691; *F, G* Bayer 2291; *H, I* Plowes 4212; *J, K* Meve 397)

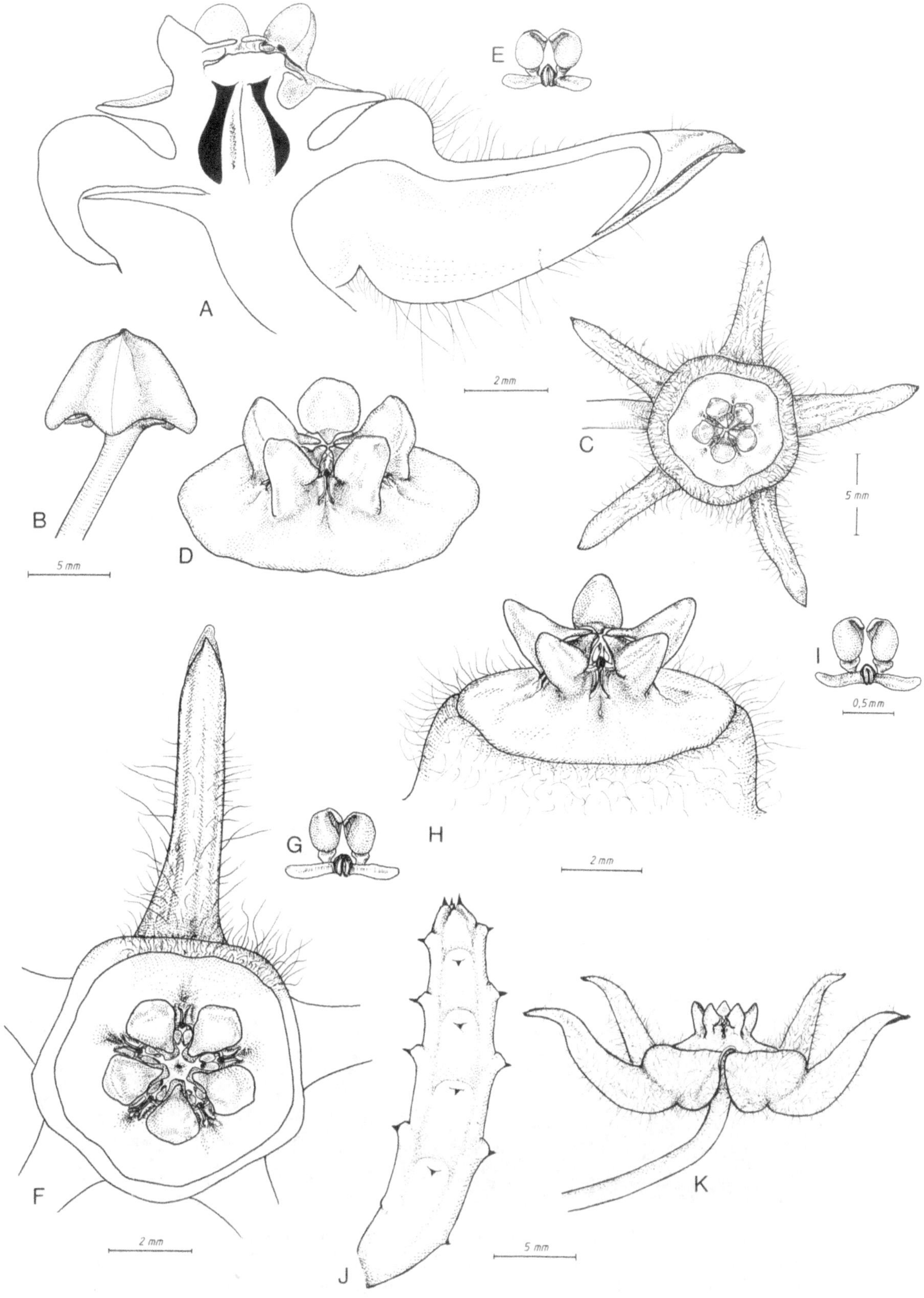
A
B
C
D
E
F
G
H
I
J
K
2 mm
5 mm
5 mm
0,5 mm
2 mm
2 mm
5 mm

Plate 3. *A Duvalia pubescens*, MEVE 153, ca. 2.5x; *B D. vestita*, HARTMANN 12691, ca. 2.5x; *C D. eilensis*, LAVRANOS, BARAD & LINDEN 24592, ca. 2.5x; *D D. galgallensis*, ex hort. DE BOER, ca. 2x (Photo: A. DE BOER); *E D. sulcata* subsp. *somalensis* (Djibouti), BUTLER D105, ca. 1.5x (Photo: W. BOSMA)

Plate 4. *A Duvalia. sulcata* subsp. *sulcata*, HARTMANN 21519, ca. 2.5x; *B D. sulcata* subsp. *seminuda*, NOLTEE 1495, ca. 2x; *C D. sulcata* subsp. *seminuda*, NOLTEE 1495, ca. 1.5x; *D D. sulcata* subsp. *somalensis*, GILBERT s.n. (Ethiopia), ca. 1.3x (Photo: M. GILBERT); *E D. velutina*, BARAD 11732, ca. 3.5x; *F D. velutina*, BARAD 11745, ca. 2.5x

vestita are widely peripatric rather than sympatric. Therefore, the strategy of *D. vestita* to avoid competition with its progenitors is obvious.

Morphologically, the distinct features of stem and flower support the independent status of *D. vestita* as well. The often elongated stems with rhizomatous tendencies, the minute leaf rudiments with almost reduced stipular glands, the stout annulus and the type of corolla indumentum allow for easy recognition of *D. vestita*. Apart from *D. vestita*, *D. polita*, and occasionally also *D. immaculata*, develop subterranean stems, and these three are the only species colonizing areas in which bush fires occur regularly.

Specimens examined

South Africa:

3319 DD: McGregor, Olifantsdoorn, BAYER 680 (NBG!, SRGH!).
3320 CC: Ashton, BAYER s.n. sub KG 299/71* (NBG!).
3419 AB: Caledon, Skuitsberg, BAYER s.n. sub KG 417/71 (NBG!, SRGH).
3419 DB: S Heuningsrug, HARTMANN 12691* (MSUN).
3419 DB: SW Bredasdorp, Heuningrug, BAYER 2260*, 2340 (NBG!); S Heuningrug, 3 km N Wiesdrif, FORRESTER 308 (NBG!).
3420 AA: 1 km E Stormsvlei Hotel, BAYER 1413* (NBG!, SRGH).
3420 AB: Swellendam, 10 km NW Malgas, LAVRANOS 28208*.
3421 AA: 1 m W Riversdale, HALL s.n. sub NBG 842/53 (NBG!); Riversdale, MUIR s.n. (NBG!).
3421 BB: N Cooper siding, BAYER 743 sub KG 324/73 (NBG).
3421 BD: 5 km inland from Gouritzmond, LAVRANOS 21170 (NBG!).
3422 AA: 8 km NE Mosselbaai, "Dumbie Dykes", BAYER 2291 et 761* (NBG!); Mosselbay, "Dumbie Dykes", BRUYNS s.n. sub KG 471/75 (NBG!).
3424 BB: Humansdorp, near Kouga Dam, STAYNER sub NBG 369/62 (NBG!).

II. *Duvalia* sect. *Arabica* MEVE & F. ALBERS, Mitt. Inst. Allg. Bot. Hamburg **23b**: 597 (1990).

Type: *Duvalia sulcata* N. E. BR.

Description. Stems globose, cylindrical to club-shaped, obscurely 4-angled, greyish-green, irregularly mottled with purplish markings, podaria subulate without clearly separated leaf rudiments, stipules absent. Inner corona lobes whitish, with an erect, acute and slightly incurved dorsal part.

Distribution. Djibouti, Ethiopia, Saudi Arabia, Somalia, Sudan, Yemen. Territories bordering the southern Red Sea and Gulf of Aden, and adjacent parts of Ethiopia and Somalia.

Key to the species

1. Apex of corolla lobes clearly hairy, lamina without longitudinal furrows . . 2
– Apex of corolla lobes glabrous or velvety, lamina with longitudinal furrows 3
2. Corolla bright mahagony, pilose all over, hairs 1.5–3 mm long, corolla lobes ± flat . . . . 15. *D. galgallensis*
– Corolla yellow or reddish-brown (or maculated), tips of corolla lobes densely pilose, lobes almost completely replicate. . . . . 14. *D. eilensis*
3. Annulus flat, 1–2 mm high, densely covered with hairs, 5–8 mm long, corolla lobes glabrous, lamina with 5 longitudinal furrows. . . . . 16a. *D. sulcata* subsp. sulcata
– Annulus fleshy, above 2 mm in height, glabrous or nearly so, lamina without, or with 3 or 5 furrows 4
4. Corolla lobe margins with clavate hairs, 2–3 mm long . . . . 16b. *D. sulcata* subsp. *seminuda*
– Corolla lobe margins without clavate hairs 5
5. Annulus glabrous, merging into the velvety haired corolla lobes . . . . 17. *D. velutina*
– Annulus occasionally papillate, otherwise corolla glabrous, brilliant. . . . . 16c. *D. sulcata* subsp. *somalensis*

14. ***Duvalia eilensis*** Lavranos, Cact. Succ. J. (US) **64**: 260 (1972).

Type: Somalia, Basoso Region (Mijertein), Eil airfield, 16 km N Eil, 08°02′ N 49°58′ E, ca. 150 m, on limestone plateau, 2. Dec. 1969, Lavranos 7223 (holo-, FT!, in spiritu, now dried out).

Description. Stems 20–40 mm long, 10–15 mm wide, 4-angled, ± ovoid to cylindrical, yellowish-green to grey-green, ± spotted with dark green or reddish-brown, glabrous, complex of podaria and leaf rudiments up to 5 mm long, conically acute. Inflorescence bostrychoid, primarily dichasial, 1- several flowers developing successively from near stem base, with excrement stench. Pedicels 20–50 mm long, 1.5–2 mm diam.; sepals 3–4 mm long, ca. 2 mm wide at the base; buds flattened with suddenly constricted apex. Corolla 20–30 mm diam., cream to bright yellow, intensely spotted with reddish-brown, rarely unspotted yellowish or uniform reddish-brown, lobes 6–8 mm long, replicate at least the apical half, ca. 3.5 mm high, basally coarsely, apically finely spotted, basal two thirds covered with hair papillae, 100–200 µm long, apical third with bent hairs up to 1.2 mm long; annulus circular, fleshy, 12–15 mm diam., 1.5–2 mm high, peripherally slowly merging into the corolla lobes, papillate, papillae up to 50 µm long, obtuse, rim of annulus centrally often with cream coloured ring. Coronal disc circular to pentagonal, 5.5–6.5 diam., cream, spotted with black-brown or completely black-brown, margin slightly undulating, cartilagineous, blackish, staminal corona lobes ivory tinged with purple, ca. 1.5 mm diam.; guide rails very short with widely gaping mouth. Pollinia ca. 400 µm long, 270 µm wide, dark yellow, pellucid margin ca. 180 µm long; caudicles ca. 150 µm long, yellow, translator wings ca. 350–400 µm long, bright yellow; corpusculum ca. 250 µm long. Follicles not seen; seeds 5 mm long, 3.5 mm wide, light brown, wing ca. 700 µm wide. Chromosome number: 2n = 22 (voucher: Lavranos & al. 24952, MSUN) (Fig. 46).

Distribution. Somalia (Fig. 47).

Habitat. Lowland species, obviously with affinity to calcareous soil (see protologue).

The biogeographically and morphologically very interesting *D. eilensis* occupies an isolated position within sect. *Arabica*. Its short and rounded staminal corona lobes resemble the ones of the typical section more than those usually found in sect *Arabica.* Taking into account this special character as well as the southernmost distribution area of *D. eilensis* within sect. *Arabica*, it seems appropriate to recognize *D. eilensis* as connecting link between both lineages. However, the similarities in corona structure are most likely of convergent nature, since the obviously complete incompatibility of *D. eilensis* with species of sect. *Duvalia* (see "Breeding system") suggest an early separation from common ancestors during the migration in two different main directions.

Specimen examined

Somalia:

Eil airstrip, 55 km from Eil to Sinugif, Lavranos, Barad & Linden 24952* (MSUN).

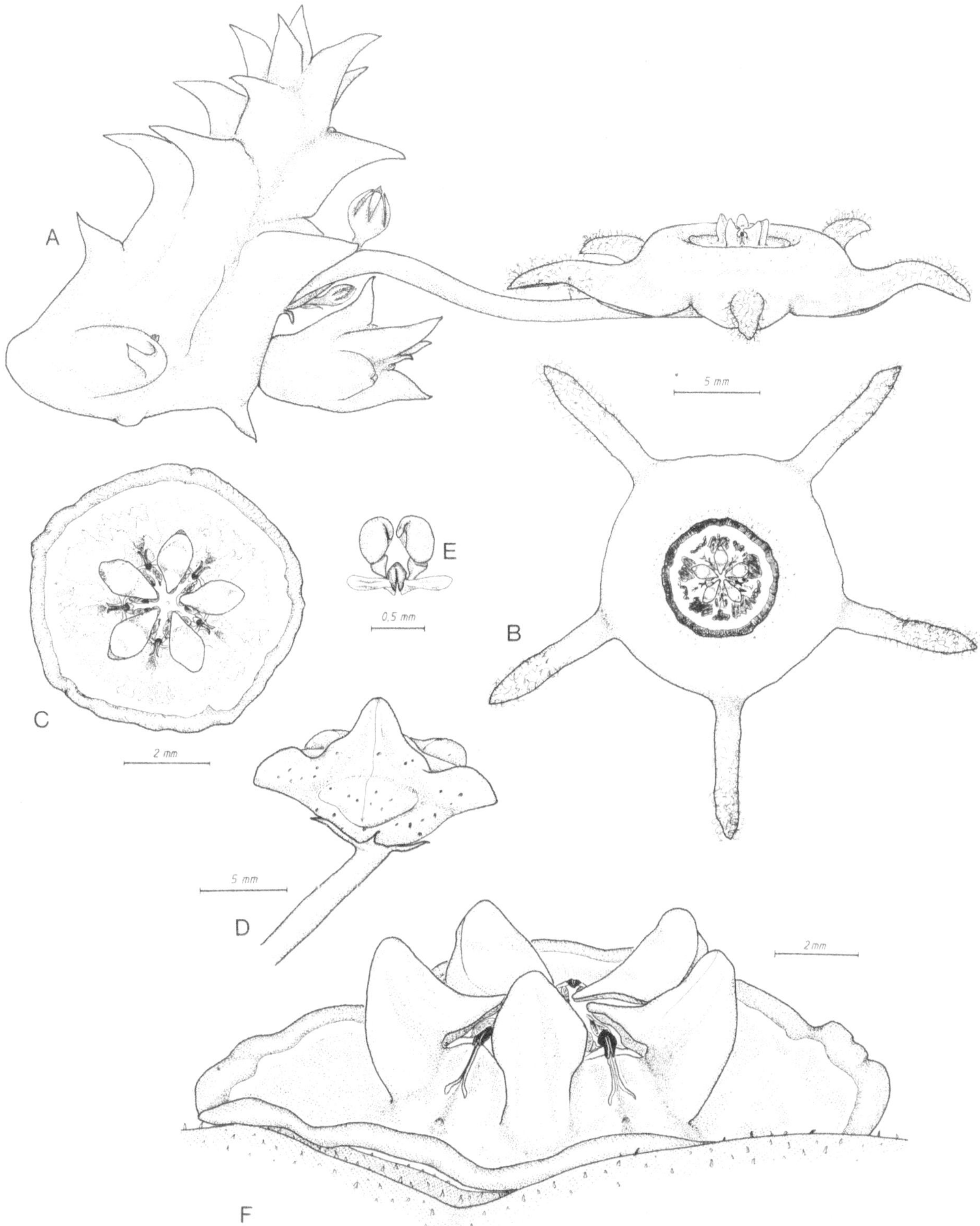

Fig. 46. *Duvalia eilensis*. *A* Flowering stem in lateral view; *B* flower in top view; *C* pollination apparatus in top view; *D* bud; *E* pollinarium; *F* pollination apparatus in lateral view. (LAVRANOS, BARAD & LINDEN 24952)

15. ***Duvalia galgallensis*** Lavranos, Cact. Succ. J. (US) **66**: 184 (1974).

Type: Somalia: Bosaso Region, rocky hilltop above the village of Galgallo, 10°59′ N 49°02′ E, ca. 1250 m a.s.l., 10. Jan. 1973, Lavranos & Horwood 10260 (holo-, FT!, in spiritu, now dried out).

Description: Stems 8–20 mm long, 7–12 mm diam., ± 4-angled, glabrous, bright yellowish-green, podarium/leaf rudiment complex ± 3 mm long, acute, hard, podaria spreading. Inflorescence dichasial or bostrychoid, 1 to several flowers developing successively from basal stem regions, with stench of excrements. Pedicels 15–20 mm long, ca. 2 mm diam.; sepals 4–5 mm long, ca. 2 mm wide at the base. Corolla 30–36 mm diam., bright mahagony-brown, brilliant; free corolla lobes 10–13 mm long, 6–7 mm wide, lamina slightly convex, margins slightly reflexed, adaxially pilose, hairs simple, up to 3 mm long, whitish or purple; annulus circular, delicate, 10–12 mm diam., ca. 1 mm high, almost completely covered by the corona disc, densely hairy, hairs ca. 1 mm long. Coronal disc ± decagonal with undulating margin, 10–11 mm diam., cream with reddish spots, staminal corona

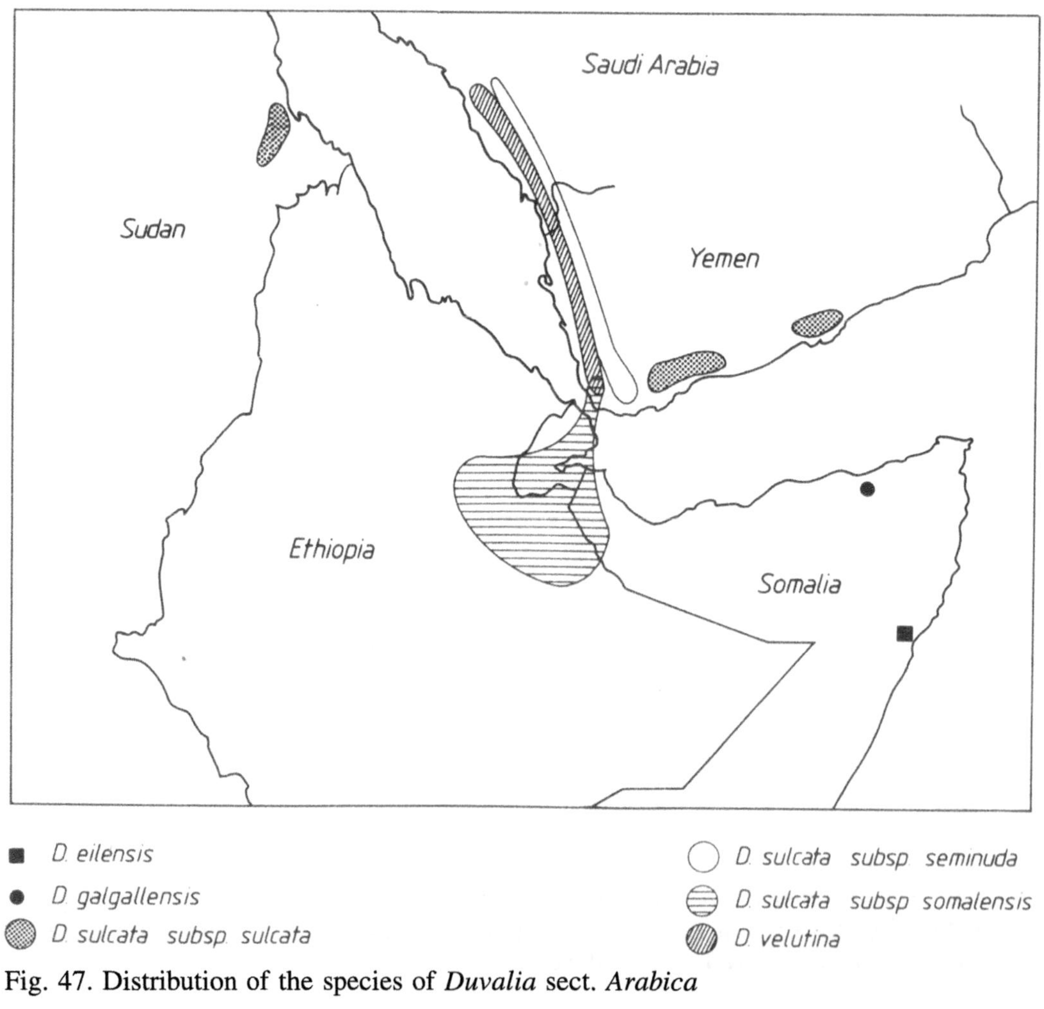

Fig. 47. Distribution of the species of *Duvalia* sect. *Arabica*

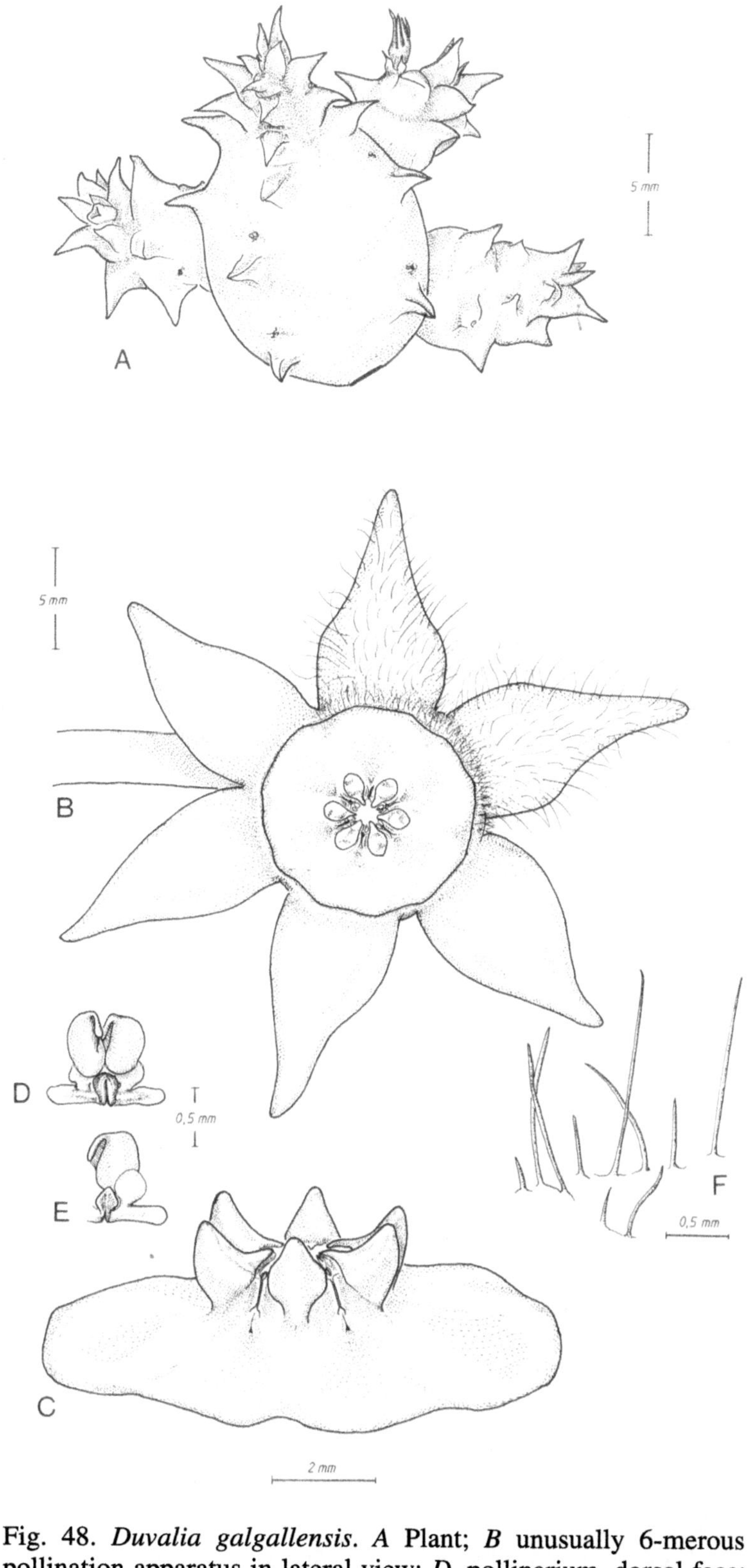

Fig. 48. *Duvalia galgallensis*. *A* Plant; *B* unusually 6-merous flower in top view; *C* pollination apparatus in lateral view; *D*, pollinarium, dorsal face; *E* single pollinium with translator, ventral face; *F* hairs of adaxial side of corolla lobe. (All s.n., ex hort. DE BOER)

lobes ca. 2 mm diam., cream, dorsal parts rounded with only slightly incurved tips. Pollinia ca. 450 µm long, 280 µm wide, dark yellow, pellucid margin ca. 200 µm long; caudicles ca. 130 µm long, translator wings ca. 400 µm long; corpusculum ca. 250 µm long. Fruits and seeds not seen. Chromosome number: $2n = 22$ (voucher: ex De Boer, MSUN) (Fig. 48).

Distribution. Somalia, see type locality (Fig. 47), no other collections known.

The flower morphology of *Duvalia galgallensis* shows similarities with *D. sulcata* (Figs. 49, 51, 52). The compact stems, the smooth and unfurrowed corolla surface with the long simple hairs and the delicate annulus are important characters illustrating the special status of this species. Like in *D. eilensis*, its isolated distribution in the Horn of Africa (Fig. 47), suggests a parallel development in space and time for both taxa. *Duvalia galgallensis* as well as *D. eilensis* should be understood as relictual, palaeoendemic species.

Specimen examined

Ex hort. DeBoer (MSUN).

16. ***Duvalia sulcata*** **N.E. Br.**, Bull. Misc. Inform. (Kew) 1910: 193 (1910).

Type: (S-)Yemen, probably from the Hadramaut, 1896/97, Bent s.n. (holo-, K!).

Key to the subspecies of *Duvalia sulcata*

1. Annulus fleshy, >2–4 mm high, glabrous, papillate or rarely pilose 2
– Annulus inconspicuous, 1–2 mm high, densely covered with white hairs, 5–8 mm long . . . . a. subsp. *sulcata*
2. Corolla lobes ± spreading, with tips often decurved, with hair papillae at the sinuses of the corolla lobes . . . . .b. subsp. *seminuda*
– Corolla lobes replicate half way down, without hair papillae at the sinuses of the corolla lobes . . . . c. subsp. *somalensis*

16a. ***Duvalia sulcata*** **subsp. *sulcata***

Description. Stems plagiotropous to nearly orthotropous, 20–80 mm long (in Sudanian plants up to 120 mm), 10–20 mm diam., ± 4-angled and club-shaped, glabrous, grey-green with dark green to reddish-brown spots, podarium/leaf rudiment complex rarely exceeding 5 mm (up to 8 mm), leaf tips soft. Inflorescence dichasial to bostrychoid, one to many prostrate flowers with intense excrement stench developing successively from basal parts of stem, synflorescence long-lived with many partial inflorescences, rachis up to 30 mm long. Pedicel 10–40 mm long, 1.5–2 mm diam., getting longer (up to 90 mm) and thicker (5 mm) when bearing follicles; sepals 3–5 mm long, 1.5–2 mm wide at the base. Corolla 30–45 mm diam., greenish-ochre, flesh-coloured, reddish-brown or bright mahagony, abaxially glabrous, lobes 10–18 mm long, 7.5–12 mm wide, lamina flat to convex with slightly recurved margins, ± brilliant, adaxially with 5 longitudinal furrows, tips finely warty, margins basally or up to the tip with clavate hairs, 1.5–2.5 mm long, whitish/purplish; annulus ± circular, rather delicate, 7–14 mm diam., 1–1.5(2) mm high, nearly covered by the coronal disc, merging with the corolla lobes, densely covered with hairs, 5–8 mm long, simple, straight, whitish or pinkish. Coronal disc 6–8 mm diam., circular, penta- to decagonal, cream to flesh-coloured, ± spotted with brown or purplish, staminal corona lobes white or ivory, 2.2–4 mm high, 1.4–2.2 mm wide, dorsal parts mostly extremely acute, erect, incurved. Pollinia 350–500 µm long, ±300 µm wide, dark yellow, pellucid margin 140–180 µm long; caudicles orange-yellow, ca. 130 µm long, translator wings 400–450 µm long; corpusculum 180–220 µm long. Follicles 12–16 cm long, 7–8 mm diam., grey-green with reddish markings; seeds ca. 6 mm long, 3–4 mm wide, grey-brown, wing ca. 1 mm broad, coma ca. 20 mm long. Chromosome number: heteroploid species, 2n = 22 (Sudan), 2n = 44 (Yemen) (vouchers: see * in "Specimens examined"). (Fig. 49).

Distribution. (S-)Yemen, Sudan, Ethiopia?; ca. 13°00′–18°45′ N and 37°00′–49°30′ E (Figs. 47, 50).

Habitat. Typical for higher altitudes: 500–2000 m a.s.l.

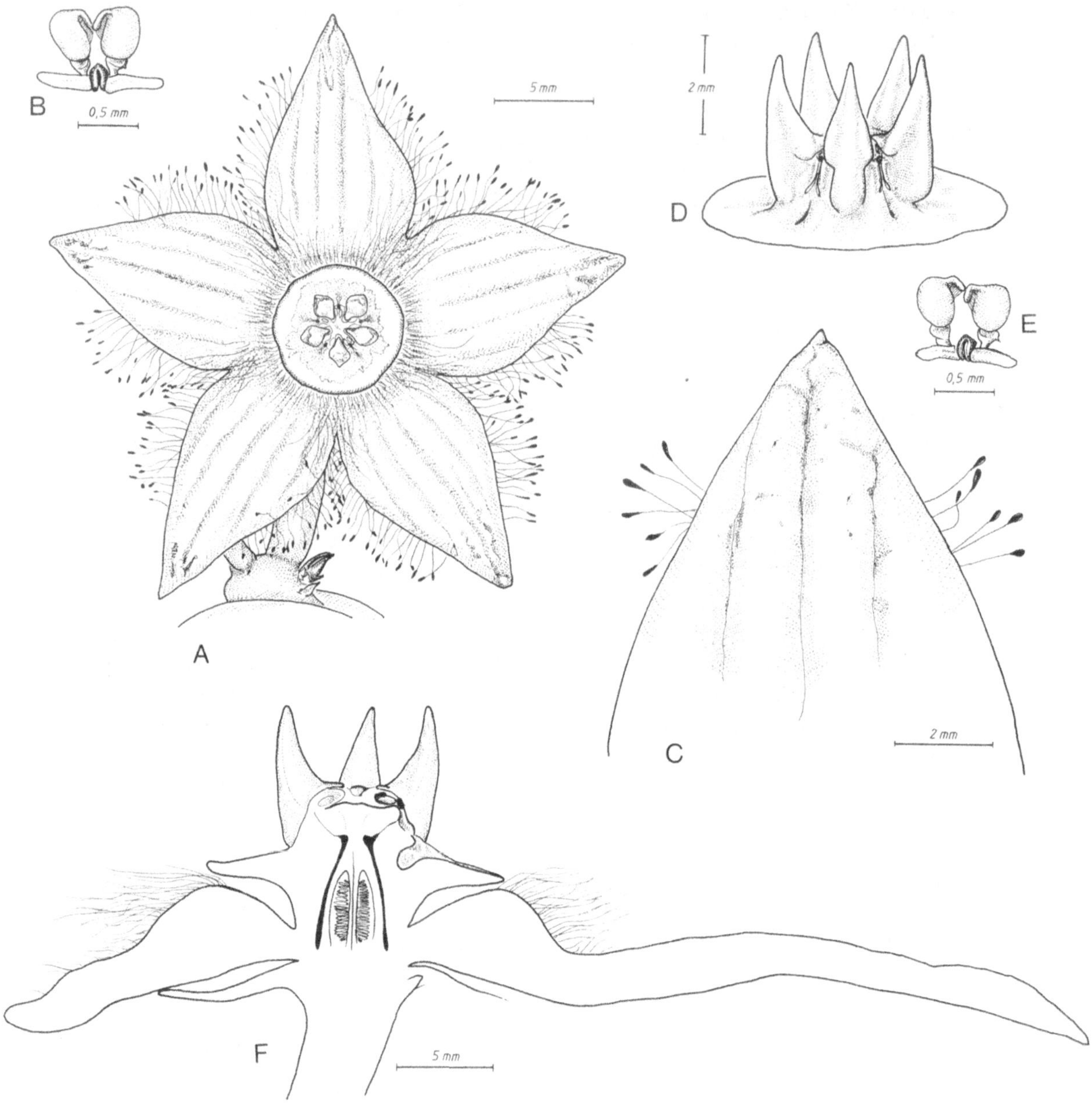

Fig. 49. *Duvalia sulcata* subsp. *sulcata*. *A* Flower in top view; *B* pollinarium; *C* tip of corolla lobe in top view; *D* pollination apparatus in lateral view; *E* pollinarium; *F* flower in longitudinal section. (*A*, *B*, *F* HARTMANN 21506; *C–E* HARTMANN 21507)

Specimens examined

Sudan:

Jebel Aulini near Erkovit, 18°45′ N/37° 04′ E, HARTMANN 21506*, 21507* (MSUN); Sinkat, 18°54′ N/36°50′ E, HARTMANN 21518*; Base of Akaba Pass, 19°01′ N/36°59′ E, HARTMANN 21519* (MSUN); Red Sea Hills, DARVALL s.n. sub Kew 1935 (K!); Amadal (100 km SW Tokar), AWAD sub PLOWES 7828* (MSUN).

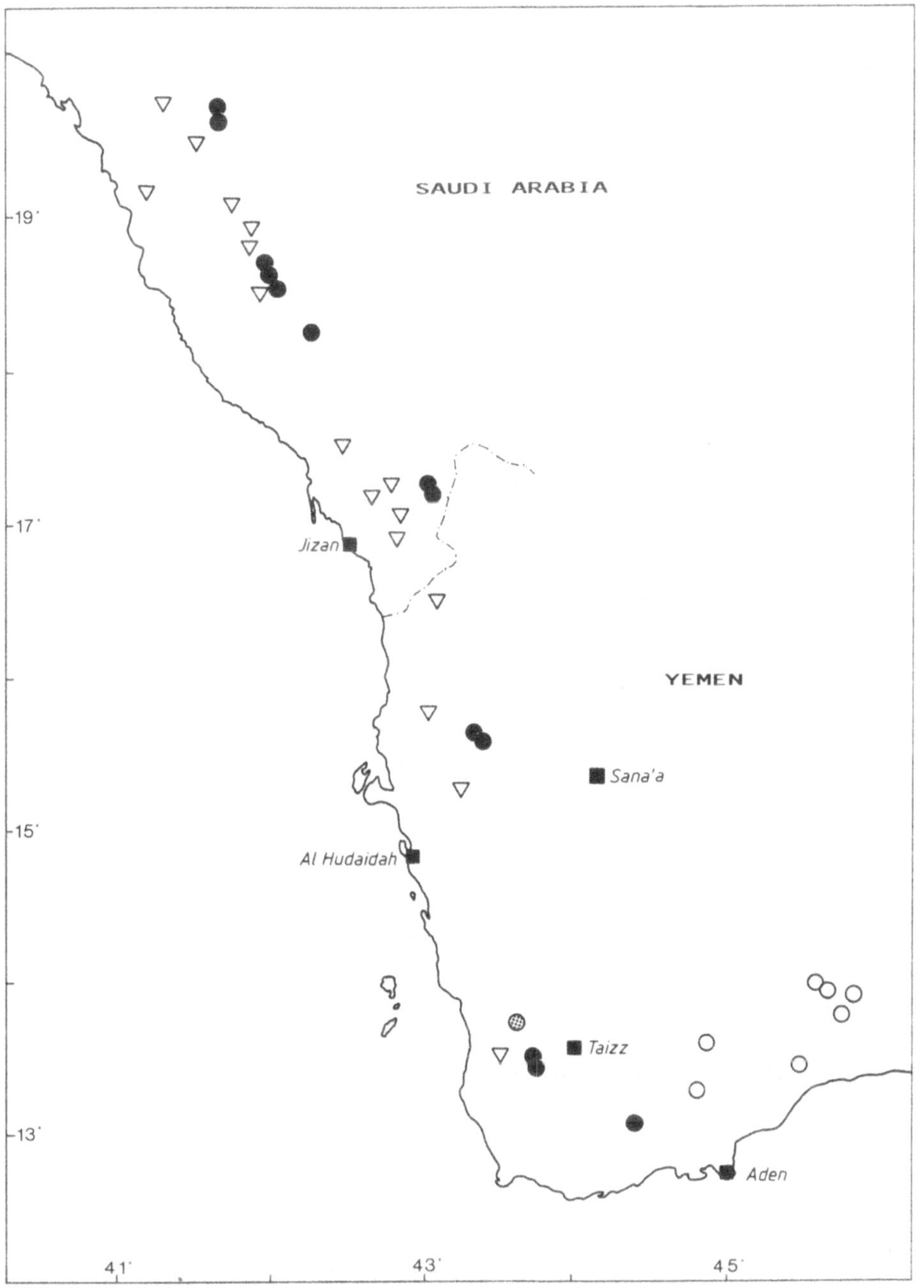

Fig. 50. Distribution of *Duvalia sulcata* subspp. *sulcata* (●), *seminuda* (○), *somalensis* (●) and *D. velutina* (▼) on the southwestern Arabian Peninsula

Yemen (S):
N Thumair, "Kutaibi Country", Lavranos 1862 (K!); Lodar, near village, Lavranos 1898; 6 mls S Lodar, Lavranos 1899 (K!); 16 km NW Mukalla (Hadramaut); Lavranos & Rauh 2925* (PRE, SRGH); ca. 50 mi NE Aden, Am Sana, Maxwell Darling 228 (K!); Lodar, Smith & Lavranos 806*; Audhali Plateau, near Aryab, Smith & Lavranos s.n. (K!).

16b. ***Duvalia sulcata* subsp. *seminuda* (Lavranos) Meve, stat. nov.** Basionym: *D. sulcata* var. *seminuda* Lavranos, Cact. Succ. J. (US) **39**(1): 5 (1967).

Type: SW Arabia, Yemen, Subaihi country, in dusty alluvial soil near Da'ar al Kudaimi, ca. 250 m a.s.l., 13°04′ N 44°23′ E, 31. March 1964, Rauh & Lavranos 3157 (holo, K!, in spiritu).

Description. Stems 20–60 mm long, 10–22 mm wide, rounded 4-angled, ± club-shaped, glabrous, grey-green with dark green to reddish-brown spots, podarium/leaf rudiment complex 5–8 mm long, suddenly merging into a soft tip. Inflorescence primarily dichasial, later bostrychoid, one to many prostrate flowers developing successively, flowers with intense stench of excrement to almost unscented; synflorescence often persistent with many partial inflorescences, rachis up to 30 mm long. Pedicels 15–40 mm long, 1.5–3 mm diam.; sepals 3–6 mm long, 1–2 mm wide at the base. Corolla 30–45 mm diam., ochre, flesh-coloured to dark reddish-brown, dull or brilliant, lobes 10–16 mm long, 7–12 mm wide, lamina slightly convex, spreading with tips often decurved, surface with 3 or 5 longitudinal furrows, glabrous but apex usually warty and with fine, white hairs, margins basally or up to the apex with clavate hairs, 2–3 mm long, whitish/purplish, with simple hair papillae, 60–300 μm long in the corolla sinusles; annulus ± circular or pentagonal, 9–12 mm diam., 2–4 mm high, massive, glabrous, pubescent or pilose, hairs up to 5 mm long, usually much shorter, white. Coronal disc 6.5–10 mm diam., ± circular, pentagonal or decagonal, at the same level as the rim of annulus, cream to flesh-coloured, patterned with reddish dots or whole rings, staminal corona lobes 2.2–4 mm high, 1.4–2.2 mm long, white, ivory or cream, rarely tinged with reddish-brown, dorsal parts blunt to acuminate, ± erect with slightly incurved tips. Pollinia 350–500 μm long, 250–300 μm wide, dark yellow, pellucid margin 160–220 μm long; caudicles ca. 130 μm long, orange-yellow, translator wings 350–450 μm long, corpusculum 200–300 μm long. Fruits and seeds not seen. Chromosome number: heteroploid with cytotypes of $2n = 22$ or $2n = 44$ (vouchers: see * in "Specimens examined") (Figs. 51, 52).

Distribution. (N-)Yemen, Saudi Arabia; 13°00′–19°45′ N and 41°45′–45°20′ E (Figs. 47, 50).

Habitat. On sandy or rocky soil, mostly in *Acacia* dominated shrubland. Typical for medium high altitudes of the western Asir and the Mountain Tihama, alt.: 250–700 m a. s. l.

Specimens examined

Saudi Arabia:

100 km NE Jizan, base of SW Jabal Fayfa, 17°14 N 43°05 E, Collenette 5545, 5871, 5912 (K!); 1 km E Muhayl on Jeddah to Gizan road, 18°32 N 42°05 E, Collenette 2708 (K!); 23 km from junction with Jeddah on Grande Lavori Rd (Muhayl to Gizan), 19°41 N 41°45 E, Collenette 2816 et 2817 (K!; ZSS); Wadi Al Uss, on western slope of Jebel Sawdah, 19°41 N 41°45 E, Collenette 2702 (K!); 1 km E Muhayl on Jeddah to Gizan road,

Fig. 51. *Duvalia sulcata* subsp. *seminuda*. *A* Flowering stem; *B* flower in lateral view; *C* annulus with pollination apparatus; *D* pollinarium; *E* clavate hair of corolla margin; *F* section of flower in top view; *G* pollination apparatus in lateral view; *H* pollinarium. (*A* Barad 11729; *B–E* Noltee 1373; *F–H* Noltee 1495)

A
5 mm
2 mm
F
5 mm
B
0,5 mm
D
C
E
2 mm
G
0,5 mm
H

Collenette 2704 (K!; ZSS); 23 km from junction with Jeddah on Grande Lavori Rd (Muhayl to Gizan), Collenette 2816 (K!; ZSS), 18 km from junction with Jeddah on Grandi Lavori Rd (Muhayl to Gizan, Collenette 3129 (K!; ZSS); Asir, 10 km N Muhayl, Lavranos & Collenette 18236*.

Yemen (N):
Al Barh, Barad 11729 (MSUN)*; Basin of At Tur (Mountain Tihama), Deil 200 (MSUN); Noltee 1373* (MSUN); 7 km W At Tur, Noltee 1495* (MSUN); 55 km W Taizz, Al Barh, Noltee 1898*, 1906* (MSUN).

16c. ***Duvalia sulcata* subsp. *somalensis* (Lavranos) Meve, comb. et stat. nov.**
Basionym: *D. somalensis* Lavranos, Cact. Succ. J. (US) **43**(2): 65 (1971).

Type: (N)Somalia, between Borama and Zeila, under low shrubs on the sandy plain around the hamlet of Abd al Qadr, 10°31′ N 42°53′ E, alt. ca. 800 m, 5. Okt 1968, Lavranos 6838 (holo-, FT!, in spiritu, now dried out).

Description. Stems 30–50 mm long, 10–15 mm wide, rounded 4-angled, ± club-shaped, glabrous, grey-green with dark green or reddish spots, podarium/leaf rudiment complex 5–7 mm long, suddenly merging into the soft leaf tip. Inflorescence primarily dichasial, later bostrychoid, 1 to several prostate flowers developing successively, with stench of excrements, synflorescence often consists of many partial inflorescences, rachis up to 20 mm long. Pedicels 25–35 mm long, 2 mm diam.; sepals ca. 3 mm long, 2 mm wide at the base. Corolla 30–35 mm diam., yellowish with reddish-brown spots to uniformly purple-brown, glabrous, ± brilliant, lobes ca. 15 mm long, 6.5–7.5 mm wide, lamina replicate ± half way down, adaxial with 3 or 5 longitudinal furrows, apex finely warty; annulus ± circular, 10–11 mm diam., 2–3 mm high, inconspicuously covered with hair papillae. Coronal disc 6.5–7.5 mm diam., ± circular to pentagonal, at the same level as the rim of annulus, yellowish with reddish-brown dots or spots or bands; staminal corona lobes 2.2–3 mm high, ca. 2 mm long, ± ivory, dorsal parts strongly acuminate, ± erect with slightly incurved tips, inner parts short, blunt or acute. Pollinia: 400–450 µm long, 250–300 µm wide, pellucid margin ca. 200 µm long; caudicles ca. 150 µm long, translator wings ca. 400 µm long; corpusculum ca. 250 µm long. Fruits and seeds not seen. Chromosome number: 2n = 22 (voucher: Butler D105, in cult. Bot. Garden Münster). (Fig. 52 F, G).

Distribution. Ethiopia, Djibouti, Yemen, Somalia; 10°00′–13°00′ N and 40°00′–43°00′ E (Figs. 47, 50).

Habitat. Poorly known, alt.: ca. 250–800 m a.s.l.

Specimens examined
Djibouti:
NW end of Barrawein, Lavranos 11253; Djibouti, Butler D105*.
Ethiopia:
Harerge Region, ca. 30 km from Dire Dawa to Djibouti, Gilbert s.n. (ETH, in spiritu).
Somalia:
65 km W Djibouti, NW Barrayer, Lavranos & Newton 13161.
Yemen (N):
5 km S Hays, Lavranos & Newton 15906 (photo only).

Fig. 52. *A–E Duvalia sulcata* subsp. *seminuda*; *A* budding stem; *B* flower in top view; *C* pollinarium; *D* centre of flower in top view; *E* pollinarium. *F*, *G D. sulcata* subsp. *somalensis*; *F* flowering stem; *G* gynostegium with staminal corona. (*A–C* Noltee 1898; *D*, *E* Barad 11729; *F*, *G* from photographs of W. Bosma taken from Butler 105D)

The typical subspecies shows an unusual disjunction for the genus. It is only found along the western side of the Red Sea and north to the Gulf of Aden. Here, the plants occur in more humid (rain and fog!) regions around 500–2100 m a.s.l. (Fig. 47). However, *Duvalia sulcata* subsp. *sulcata* is presumably also an element of the Ethiopian Flora (M. GILBERT, pers. comm.). Most strikingly, the flat annulus possesses a dense cover of shiny white, straight hairs (Fig. 49F). Also, the adaxial side of the corolla lobes is totally glabrous, and the staminal corona lobes are longer and more acute than in the other subspecies (Fig. 49D). There is little infrasubspecific morphological variation. However, the stems of the African plants are more robust; and the corolla of the African forms seen are usually ochre, while the Arabian ones tend to be flesh-coloured or of reddish-brown. Karyologically, there are two cytotypes, the diploid Sudanian one and the tetraploid Arabian one.

Duvalia sulcata subsp. *seminuda* (Figs. 51, 52) is restricted to southwestern Saudi Arabia and the western parts of the Yemen at altitudes between 150 and 800 m (Figs. 47, 50). This taxon is unmistakably marked by the small pubescent area restricted to the sinuses of the corolla lobes (Fig. 51F). Fine hairs can be also found on the tips of the corolla lobes. Also, the clavate hairs along the margins of the corolla were found to be longer than in the typical subspecies. Less subtle and cryptic, the mighty and high annulus is the best marker for this taxon, which is glabrous or only sparsely indumented (Fig. 51C). Glabrous to long-hairy individuals occur in mixed populations in Saudi Arabia (S. COLLENETTE, pers. comm.).

It can be assumed that *D. sulcata* subsp. *seminuda* has been distributed parallel to the Red Sea, along the climatically favoured mountain ranges of SW Arabia as far as 20° northern latitude.

In the higher altitudes of SW Yemen, *D. sulcata* subsp. *seminuda* occurs possibly parapatric with tetraploid populations of the typical subspecies, however, there are no hints to a hybrid zone. Common localities of *D. sulcata* subsp. *seminuda* and *D. velutina* are known as well (SW N-Yemen: Al Barh; S Saudi Arabia), even though the two taxa are usually separated precisely because of their ecological preferences. Hybrids, again, are not reported. Potential bastards of the two taxa have been found in other regions (see: Specimens examined).

Duvalia sulcata subsp. *somalensis* colonizes altitudes between 250 and 800 m a.s.l., comparable to conditions preferred by subsp. *seminuda* as well. The almost glabrous, considerably brilliant corolla with widely reflexed lobes and the colour pattern make the best diagnostic features, separating *D. sulcata* subsp. *somalensis* from the other subspecies.

A far-reaching vicariance of the three taxa stands for their successful speciation. Only at the southwesternmost point of the Arabian Peninsula are all three *D. sulcata* taxa as well as *D. velutina* in close contact (Fig. 47).

17. ***Duvalia velutina*** **Lavranos**, Cact. Succ. J. (US) **55**(1): 24 (1983).

Type: Saudi Arabia, Jizan Province, Abu Arish, near the Agricultural Experimental Station on old lava flow in aeolian silt under bushes, ca. 100 m a.s.l., 16°59′ N 42°51′ E, Lavranos & Collenette 18243 (holo, E; iso-, MO?, not found).

Description. Stems 25–60 (−90) mm long, 10–20 mm wide, 4-(5-)angled, club-shaped, glabrous, grey-green with dark green to reddish-brown spots, podarium/leaf rudiment complexes 5–11 mm long. Inflorescence primarily dichasial, later bostrychoid, one to several flowers with carrion stench developing successively, inflorescence axis long-lived, rachis up to 30 mm long. Pedicels 30–100 mm long, 1–2 mm diam.; sepals ca. 5 mm long, basally 2–3 mm wide. Corolla 20–45 mm diam., green, flesh-coloured to reddish-brown, lobes 10–16 mm long, 10–12 mm wide, lamina ± flatly spreading, convex to slightly replicate, adaxially with 5–7 longitudinal furrows, originating along the lower rim of annulus, corolla rarely without furrows, velvety all over, hairs 100–150 µm long, apically up to 400 µm long and usually verrucose, corolla margins glabrous or with hairs along basal half or less, hairs simple, up to 1 mm long, whitish; annulus ± circular to pentagonal, merging into the corolla lobes, convex, rarely flattened, upper rim brighter in colour than the other parts of the corolla, glabrous or papillate, papillae up to 80 µm long. Coronal disc 7–8.5 mm diam., ± circular or pentagonal, at the same level as the rim of annulus, cream- to flesh-coloured with flesh-coloured to reddish-brown dots, spots or whole circles, staminal corona lobes 2.5–5 mm high, 1–2 mm long, white to ivory, outer parts vertical to spreading outwards, acuminate with in- or outwardly bent tips, inner parts blunt or acute. Pollinia 400–450 µm long, 240–280 µm wide, dark yellow to orange-brown, pellucid margin 150–200 µm long; caudicles ca. 150 µm long, 100–200 µm wide, orange-yellow, translator wings 320–400 µm long; corpusculum 200–250 µm long. Follicles ca. 10 cm long, 12 mm wide (cf. protologue, Lavranos 1983); seeds 6 mm long, 4 mm wide, brown, wing ca. 800 µm wide, light brown. Chromosome number: 2n = 22 (vouchers: see * in "Specimens examined") (Figs. 53, 54).

Habitat. A character species of the lower regions of the Tihama, alt: 50–400 m a.s.l.

Distribution. Saudi Arabia, Yemen; 13°00′-19°50′ N and 41°15′–43°00′ E (Figs. 47, 50).

Duvalia velutina has to be regarded as the most advanced species of sect. *Arabica*. The rather pyramidal habit of the flowers is unparalleled in *Duvalia* (Figs. 53A, 54D), and pedicels reaching lengths of around 7 cm (Fig. 53A) are only rarely found in other stapeliads, e.g. in *Tromotriche longipes* (A.C. Lückh.) Bruyns.

The distribution map shows the distinct eco-geographical habitat preference. *Duvalia velutina* is restricted to the coastal lowland of the western Arabian Peninsula and the lower regions of the Mountain Tihama. Here, much less rainfall occurs (80–200 mm p. a.), and it is much hotter than at the higher altitudes preferred by *D. sulcata* subsp. *seminuda*. However, the coastal areas are known for higher air humidity (Deil & Müller-Hohenstein 1988). Deil (1988b, and pers. comm.) confirmed this, reporting the geobotanical differentiation of the two species in consequence to the climatic regime. This vicariance, therefore, is probably responsible for the sparse occurrence of hybridisation (Fig. 54I-K) despite the partial crossability demonstrated in this study.

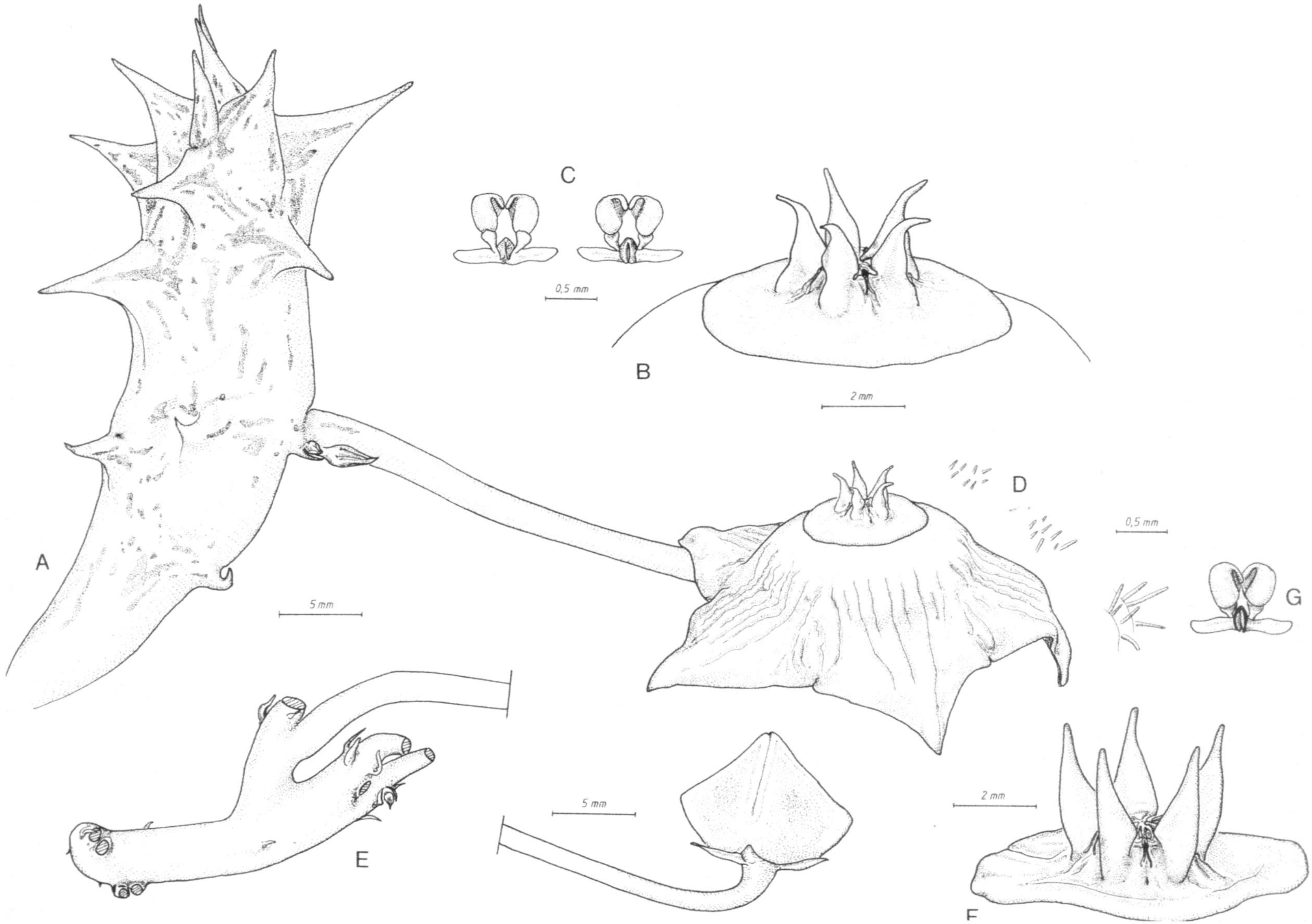

Fig. 53. *Duvalia velutina*. *A* Flowering stem; *B* pollination apparatus; *C* pollinaria, left ventral face, right dorsal face; *D* corolla indumentum of the annulus (left), basal corolla lobe (medium) and apical corolla lobe (right); *E* synflorescence; *F* pollination apparatus in lateral view; *G* pollinarium. (*A–D* Barad 11723; *E* Deil 1240; *F*, *G* Collenette 2726)

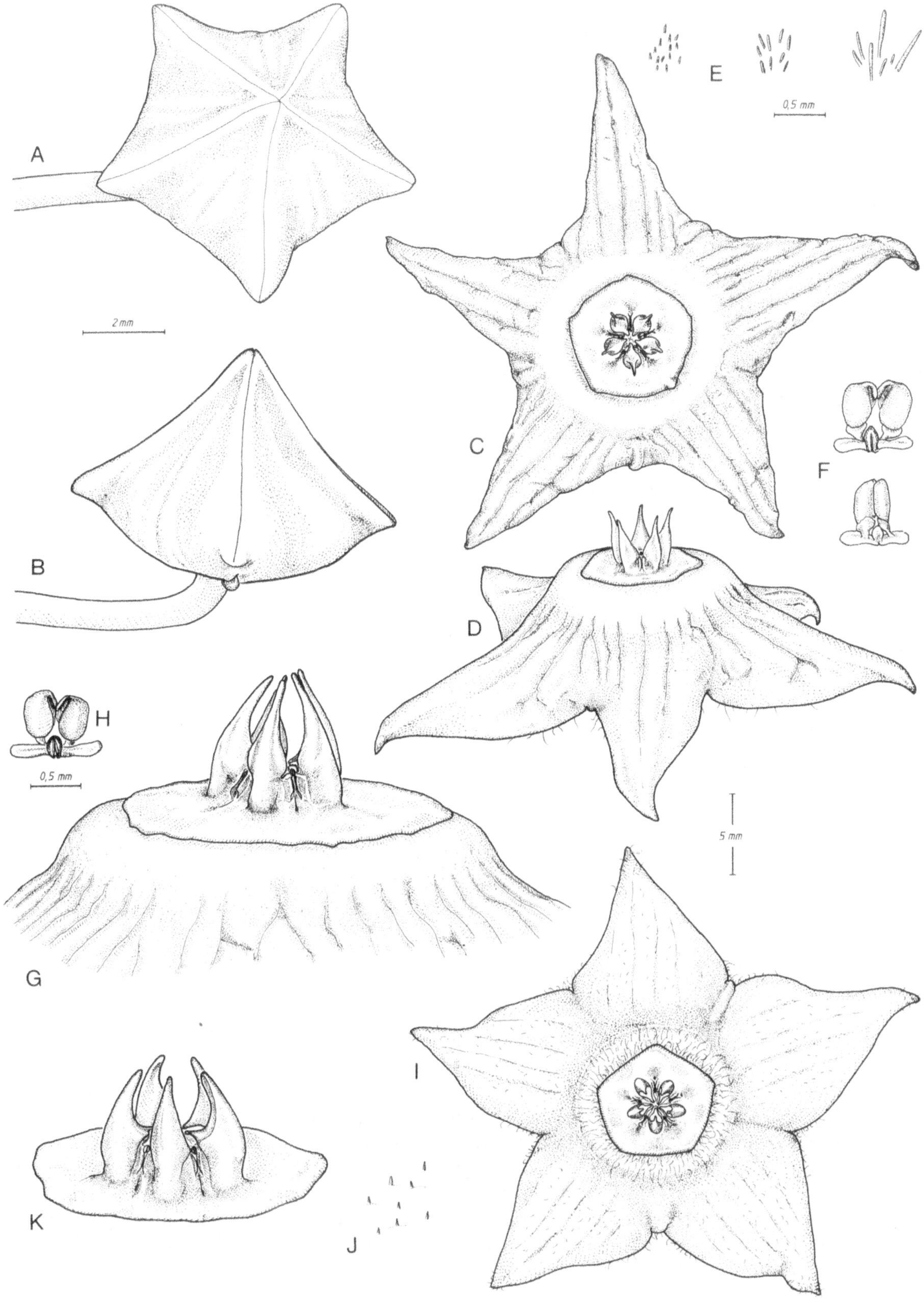
A
B
C
D
E
F
G
H
I
J
K
2 mm
0,5 mm
0,5 mm
5 mm

Specimens examined
Saudi Arabia:
To the NW of AdDarb-Baysh road, near Jizan, 17°30 N 42°30 E, COLLENETTE 5604 (K!); 1 km E Muhayl on Jeddah to Gizan road, at base of Wadi Al Uss, 18°32 N 42°05 E, COLLENETTE 2708 (K!, ZSS!); Jeddah to Gizan Road, N of Junction with Al Quiz road, 19°12 N 41°55 E, COLLENETTE 2715* (K!); 482 km from Jeddah on the Muhayl-Gizan Road, 19°20 N 41°55 E, COLLENETTE 2215 (K!); 100 km N Mukaylah and 25 km junction with Grandi Lavori road, 19°30 N 41°40 E, COLLENETTE 3945 (K!); ca. 20 km E Qunjudhah, Jebel Mershid, COLLENETTE & HIGGINBOTTOM 2726 (K!, ZSS!).
Yemen (N):
6 km N Bafil, BARAD 11723* (MSUN); 5 km N Mafrag al Hodeida, BARAD 11745* (MSUN); border of village near Zurah (Mountain Tihamah), DEIL 1240 (MSUN); near Haradh, WOOD 2682 (K!).
Hybrids seen: *Duvalia sulcata* subsp. *seminuda* × *D. velutina*
Yemen (N):
N Jamaia (N Haradh), Wadi Melah, WOOD 3055 (K!).
Saudi Arabia:
Between Sayyabah and Jabal Fayfa, COLLENETTE 3741 (K!, ZSS!).

Natural intergeneric hybrid: × ***Duvaliaranthus*** **BRUYNS**, S. Afr. J. Bot. **42**: 365 (1976). **Type** (and only) **species:** ***D. albostriatus*** BRUYNS, J. S. Afr. Bot. **42:** 365 (1976).

Type: South Africa, Cape Prov., Farm Vierfontein, 17 miles NE Vanrhynsdorp on road to Nieuwoudtville, BRUYNS 91/75 (holo-, NBG).

Conservation status

Most *Duvalia* species are common in their particular veld types, but some differentiation is necessary. *Duvalia caespitosa, D. polita* and *D. pubescens* are the most abundant species. *Duvalia maculata* is not rare considering distributional data, but it is potentially threatened because it is rarely found in mass-stands. This applies also to *D. parviflora*. However, a few species, especially *D. pillansii* (South Africa) as well as *D. galgallensis* and *D. eilensis* (Somalia), should be treated as rare and threatened. All these species have a very restricted distribution and have been collected only once or a few times.

Nomina dubia

Duvalia anemoniflora **(DEFLERS) R.A. DYER & LAVRANOS**, Fl. Pl. Afr. **44**(174), sub t. 1734 (1977). Basionym: *Stapelia anemoniflora* DEFLERS, Bull. Soc. Bot. France

---

Fig. 54. *A–G Duvalia velutina*; *A* bud in top view; *B* bud in lateral view; *C* flower in top view; *D* flower in lateral view; *E* corolla indumentum of the annulus (left), basal corolla lobe (medium) and apical corolla lobe (right); *F* pollinaria, dorsal face (above), ventral face (below); *G* annulus with pollination apparatus in lateral view; *H* pollinarium. *I–K* Putative hybrid between *D. velutina* × *D. sulcata*; *I* flower in top view; *J* hair papillae of adaxial side of corolla lobe; *K* pollination apparatus in lateral view. (*A–F* COLLENETTE 3615; *G, H* BARAD 11745; *I–K* COLLENETTE 3741)

43: 118 (1896); *Caralluma anemoniflora* (DEFLERS) A. BERGER, Stap. und Kleinien, 125 (1910).

Type: Yemen, Bilad Fodhli (östlicher Teil des Hinterlandes von Aden), 150–200 m, March 1890, DEFLERS 387 [lecto-, Icon. DEFLERS, Asclép. Arab. Trop., t. 6, designated by DYER (1977)].

Nothing of DEFLERS' type material could be traced, and all information must be gathered from his description and the plate (DEFLERS 1896: t. 6). The assumption of MEVE & ALBERS (1990b) that this taxon was described after a bud was manually opened, and was therefore showing an undeveloped, atypical corona, is to be rejected. The dimensions of the flower given by DEFLERS (corolla lobes 30 mm long, corona up to 6 mm high) are definitely too large for *D. sulcata* or any other *Duvalia*. The corona shows similarities rather with *Huernia penzigii* N.E. BR.; however, the deeply incised and furrowed corolla excludes it from *Huernia*. DYER (1977) reported on LAVRANOS' visit to DEFLERS' type locality, where he found only *D. sulcata*. I now consider *Stapelia anemoniflora* most likely representing an inter-generic hybrid of unknown parentage, which is most likely extinct today.

The following taxa are all based on cultivars. Due to the lack of any type material they must be marked as insufficiently known. However, it is most likely that they are synonyms of *D. caespitosa* var. *caespitosa*:

Duvalia concolor (SALM-DYCK) SCHLTR., J. Bot **36**: 477 (1898). Basionym: *Stapelia concolor* SALM-DYCK, Hort. Dyck., 372 (1834).

Duvalia glomerata HAW., Syn. Pl. Succ., 46 (1812). Basionym: *Stapelia glomerata* HAW., Syn. Pl. Succ., 46 (1812).

Duvalia laevigata HAW., Syn. Pl. Succ., 46 (1812). Basionym: *Stapelia laevigata* HAW., Syn. Pl. Succ., 46 (1812).

Duvalia tuberculata HAW., Syn. Pl. Succ., 46 (1812). Basionym: *Stapelia tuberculata* HAW., Syn. Pl. Succ., 46 (1812).

Stapelia barbata SALM-DYCK, *nomen nudum*, Hort. Dyck., 372 (1834).

Stapelia cymosa SCHULT. in H. ROEMER & SCHULT., Syst. Veg. **6**: 49 (1820).

Excluded taxa

Duvalia pedunculata (MASSON) LOUD. = *Tromotriche pedunculata* (MASSON) HAW.
Duvalia deflexa (JACQ.) LOUD. = *Stapelia deflexa* JACQ.
Duvalia serrulata (JACQ.) LOUD. = *Piaranthus decorus* (MASSON) N.E. BR. subsp. *decorus*
Duvalia procumbens R.A. DYER = *Huernia procumbens* (R.A. DYER) L.C. LEACH
Duvalia tanganyikensis E.A. BRUCE & P.R.O. BALLY = *Huernia tanganyikensis* (E.A. BRUCE & E.A. BALLY) L.C. LEACH
Duvalia andreaeana RAUH = *Huernia andreaeana* (RAUH) L.C. LEACH

References

ACOCKS, J. P. H., 1988: Veld Types of South Africa. – Mem. Bot. Surv. S. Africa **57:** 1–146.

ALBERS, F., 1976: *Asclepiadaceae.* – In LÖVE, A., (Ed.): IOPB chromosome number reports LI. – Taxon **25:** 161–163.

– 1977: *Asclepiadaceae.* – In LÖVE, A., (Ed.): IOPB chromosome number reports LVIII. – Taxon **26:** 557–558.

– 1983: Cytotaxonomic studies in African *Asclepiadaceae.* – Bothalia **14:** 795–798.

– MEVE, U., 1991: Mixoploidy and cytotypes. A study of possible vegetative species differentiation in stapeliads, *Asclepiadaceae.* – Bothalia **21:** 67–72.

– LIEDE, S., MEVE, U., 1993: Deviating Chromosome numbers in *Asclepiadaceae.* – Nordic J. Bot. **13:** 37–39.

– DELFS, W., KUSCH, G., MEVE, U., 1989: Lebensformen der *Ceropegieae* und *Stapelieae* in ariden Zonen Afrikas. – Beitr. Biol. Pflanzen **64:** 59–74.

AL-HUBAISHI, A., MÜLLER–HOHENSTEIN, K., 1984: An introduction to the vegetation of Yemen. – Eschborn: Deutsche Gesellschaft für Technische Zusammenarbeit (GTZ).

BACHMANN, K., 1995: Progress and pitfalls in systematics: cladistics, DNA and morphology. – Acta Bot. Neerl. **44:** 403–419.

BALLY, P. R. O., HORWOORD, F. K., LAVRANOS, J. J., 1975: A monograph of the genera *Pseudolithos* and *Whitesloanea.* – Natl. Cact. Succ. J. **30:** 31–36, 88–93.

BARAD, G. S., 1990: Pollination of the stapeliads. – Cact. Succ. J. (Los Angeles) **62:** 130–140.

BAYER, M. B., 1977: The Cape Duvalias. – Bull. African Succ. Pl. Soc. **12:** 25–27.

– 1984a: Duvalias. – Veld Fl. (1975+) **70:** 11–14.

– 1984b: The Cape Flora and the Karoo – a winter rainfall biome versus a Fynbos biome. – Veld Fl. (1975+) **70:** 17–19.

– HAROLD, K., 1987: Artificial hybridisation in *Duvalia.* – Asklepios **41:** 25–31.

BERGER, A., 1910: Stapelieen und Kleinien. – Stuttgart: Ulmer.

BOELE, C., KROESEN, A. C. J., NOLTEE, F. K. A., 1987: Checklist of *Brachystelma, Ceropegia, Riocreuxia* and the *Stapelieae.* – West Sussex: Southern Reprographics.

– 1990: Supplement to the checklist of *Brachystelma, Ceropegia, Riocreuxia* and the *Stapelieae.* – West Sussex: Southern Reprographics.

BOND, P., GOLDBLATT, P., 1984: Plants of the Cape Flora – a descriptive catalogue. J. S. African Bot., Suppl. **13:** 1–455.

BRAUN-BLANQUET, J., 1964: Pflanzensoziologie, 3rd edn. – Wien: Springer.

BROWN, N. E., 1876: *Duvalia polita*, n. sp. – Gard. Chron. **6:** 130.

– 1883: *Duvalia angustiloba* sp. nov. – Gard. Chron. **20:** 230.

– 1908: *Asclepiadaceae.* – In THISELTON-DYER, W. T., (Ed.): Flora Capensis **4**(1), pp. 518–1036. – London: Reeve.

BROWN, R., 1810: On the *Asclepiadeae*, a natural order of plants separated from the *Apocinae* of Jussieu. – Mem. Wern. Soc. **1:** 12–58. – Edinburgh.

BRUMMITT, R. K., POWELL, C. E., (Eds), 1992: Authors of plant names. – Royal Botanic Garden, Kew.

BRUYNS, P. V., 1976: A putative inter-generic stapeliad hybrid (*Asclepiadaceae*). – J. S. African Bot. **42:** 365–367.

– 1981a: Stapeliads of the Pretoria area. – Asclepiadaceae **22:** 3–10.
– 1981b: Field hybridisation. A further comment. – Asklepios **24:** 53–54.
– 1981c: A review of *Pectinaria* HAW., *Stapeliopsis* PILLANS and a new genus *Ophionella* (*Asclepiadaceae*). – Cact. Succ. J. Gr. Brit. **43:** 61–83.
– 1983: Resurrection of *Quaqua* N. E. BROWN (*Asclepiadaceae-Stapelieae*) with a critical review of the species. – Bradleya **1:** 33–78.
– 1985: Notes on Ceropegias of the Cape Province. – Bradleya **3:** 1–47.
– 1987: The genus *Caralluma* R. BR. (*Asclepiadaceae*) in Israel. – Israel J. Bot. **36:** 73–86.
– 1993: A revision of *Hoodia* and *Lavrania* (*Asclepiadaceae-Stapelieae*). – Bot. Jahrb. Syst. **115:** 145–270.
– 1995: A re-assessment of the genera *Tridentea* HAW. and *Tromotriche* HAW. – S. African J. Bot. **61:** 180–208.
– FORSTER, P. I., 1991: Recircumscription of the *Stapelieae* (*Asclepiadaceae*). – Taxon **40:** 381–391.
– MEVE, U., 1995: The generic position of *Caralluma dodsoniana*. – Edinburgh J. Bot. **52:** 195–203.

COLLENETTE, S., 1985: An illustrated guide to the flowers of Saudi Arabia. – Kingdom of Saudi Arabia Flora Publication No. **1.** – London: Scorpion.

COURT, D., 1981: Succulent flora of southern Africa. – Cape Town: Balkema.

DANNENBAUM, C., SCHILL, R., 1991: Die Entwicklung der Pollentetraden und Pollinien bei den Asclepiadaceen. – Biblioth. Bot. **141.**

DECAISNE, J., 1844: *Asclepiadeae.* In DE CANDOLLE, A. P.: Prodromus systematis naturalis regni vegetabilis **8**, pp. 490–664. – Paris: Fortin & Masson.

DEFLERS, M. A., 1896: Descriptions de quelques plantes nouvelles ou peu connues de l'Arabie meridionale II. –Bull. Soc. Bot. France **43:** 104–123.

DEIL, U., 1988a: Lebende Schätze des Jemen. – Jemen-Report **19:** 7–13.
– 1988b: Primäre und sekundäre Standorte sukkulentenreicher Pflanzengesellschaften in Südwest–Arabien. – Flora **180:** 41–57.
– MÜLLER–HOHENSTEIN, K., 1988: Euphorbias from 'Arabia Felix'. Habitats and distribution. – Euphorbia J. **5:** 108–120.

DEMETER, K., 1922: Vergleichende Asclepiadaceen–Studien. – Flora **115:** 130–175.

DYER, R. A., 1969: New and interesting records of African plants. – Bothalia **10:** 363–371.
– 1977: *Duvalia sulcata.* – Fl. Pl. Africa **44:** t. 1734.

EDWARDS, D., LEISTNER, O. A., 1971: A degree reference system for citing biological records in Southern Africa. – Mitt. Bot. Staatssamml. München **10:** 501–509.

EHRENDORFER, F., 1962: Beiträge zur Phylogenie der Gattung *Knautia* (*Dipsacaceae*). I. Cytogenetische Grundlagen und allgemeine Hinweise. – Oesterr. Bot. Z. **109:** 276–343.
– 1980: Polyploidy and distribution. – In LEWIS, W. H., (Ed.): Polyploidy, biological relevance, pp. 45–60. – New York: Plenum Press.
– 1991: Evolution und Systematik. – In SITTE, P., ZIEGLER, H., EHRENDORFER, F., BRESINSKY, A., (Eds): Lehrbuch der Botanik für Hochschulen, 33rd edn., pp. 471–828. – Heidelberg: Fischer.

FICI, S., 1991: Floristic relations between eastern Africa and the Mediterranean region with special references to northern Somalia. – Fl. Medit. **1:** 175–185.

GOLDBLATT, P., 1978: An analysis of the flora of southern Africa: its characteristics, relationships and origins. – Ann. Missouri Bot. Gard. **65:** 369–436.

HARBORNE, J. B., TURNER, B. L., 1984: Plant chemosystematics. – London: Academic Press.

HAWORTH, A. H., 1812: Synopsis plantarum succulentarum. – London: Taylor.

HEGNAUER, R., 1989: *Asclepiadaceae.* – In HEGNAUER, R. (Ed.): Die Chemotaxonomie der Pflanzen **8**, pp. 84–95. – Stuttgart: Birkhäuser.

HOFMANN, U., SPECHT, A. K., 1986: Der morphologische Charakter der Asclepiadaceencorona. – Beitr. Biol. Pflanzen **61:** 79–85.

HOOKER, J. D., 1874: *Stapelia corderoyi.* – Bot. Mag., t. 6082.

HOWIE, G. L., 1991: *Sarcostemma australe*: delayed fruit development, a unique characteristic. – S. Austral. Naturalist **65:** 74–76.

HUBER, H., 1967: 114. *Asclepiadaceae.* – In MERXMÜLLER, H., (Ed.): Prodromus einer Flora von Südwestafrika. – Lehre: Cramer.

JACQUIN, N. J., 1806–1819: Stapeliarium in hortus Vindobonensibus cultarum. 64 plates. – Wien.

JAHNKE, G., 1989: Untersuchungen zur Bestäubungsbiologie verschiedener Arten der Tribus *Stapelieae* aus der Hellskloof (Richtersveld), Südafrika. – Staatsexamensarbeit, University of Münster.

JÜRGENS, N., 1990: Remarks on the biogeography of *Mesembryanthemaceae* and their probable use in other arid regions. – Proceedings of the 12th meeting of the Association for the taxonomic study of the Flora of Tropical Africa, AETFAT. – Mitt. Inst. Allg. Bot. Hamburg **23b:** 1047–1060.

– 1991: A new approach to the Namib Region. Part I: Phytogeographic subdivision. – Vegetatio **97:** 21–38.

KASSAS, M., 1956: The mist oasis of Erkwit, Sudan. – J. Ecol. **44:** 180–194.

KNAAK, L. E., COLLINS, R. P., CHANG, N., 1969: Anthocyanins of *Stapelia.* – Lloydia **32:** 148–152.

KÖNIG, P., 1987: Vegetation und Flora im südwestlichen Saudi-Arabien (Asir, Tihama). – Diss. Bot. **101:** 1–257. – Berlin: Cramer.

KUBITZKI, K., 1967: Laubblattflavonoide als systematisches Merkmal. – Ber. Deutsch. Bot. Ges. **80:** 757–763.

KUNZE, H., 1982: Morphogenese und Synorganisation des Bestäubungsapparates einiger Asclepiadaceen. – Beitr. Biol. Pflanzen **56:** 133–170.

– 1990: Morphology and evolution of the corona in *Asclepiadaceae* and related families. – Trop. Subtrop. Pflanzenwelt **76.** – Wiesbaden: Steiner.

– 1994: Ontogeny of the translator in *Asclepiadaceae* s. str. – Pl. Syst. Evol. **193:** 223–242.

KÜRSCHNER, H., 1986: Omanisch-makranische Disjunktionen. Ein Beitrag zur pflanzengeographischen Stellung und zu den florengenetischen Beziehungen Omans. – Bot. Jahrb. Syst. **106:** 541–562.

LEACH, L. C., 1969: *Stapelieae* from South Tropical Africa, V. – Bothalia **10:** 45–54.

– 1978: A contribution towards a new classification of *Stapelieae* (*Asclepiadaceae*) with a preliminary review of *Orbea* HAW. and descriptions of three new genera. – Excelsa, Taxon. Ser. **1:** 1–75.

– 1980: A review of *Tridentea* HAW. (*Asclepiadaceae*). – Excelsa, Taxon. Ser. **2:** 1–68.

– 1984: A revision of *Tromotriche* HAW. (*Asclepiadaceae*). – J. S. African Bot. **50:** 549–562.

– 1985: A revision of *Stapelia* L. – Excelsa, Taxon. Ser. **3:** 1–157.

– 1988: A revision of *Huernia* R. BR. (*Asclepiadaceae*). – Excelsa, Taxon. Ser. **4:** 1–197.

LEWIS, W. H., 1980: Polyploidy in species populations. – In LEWIS, W. H., (Ed.): Polyploidy – biological relevance, pp. 103–144. – New York: Plenum Press.

LIEDE, S., KUNZE, H., 1993: A descriptive system for corona analysis in *Asclepiadaceae* and *Periplocaceae*. – Pl. Syst. Evol. **185:** 275–284.

– WEBERLING, F., 1995: On the inflorescence structure of *Asclepiadaceae*. – Pl. Syst. Evol. **197:** 99–109.

LÖVE, A., 1951: Taxonomical evaluation of polyploids. – Caryologia **3:** 263–284.

LOUDON, J. C., 1841: An encyclopaedia of plants. – London: Longman.

MABRY, T. J., MARKHAM, K. R., THOMAS, M. B., 1970: The systematic identification of flavonoids. – New York, Heidelberg, Berlin: Springer.

MARLOTH, R., 1908: Das Kapland, insbesonderheit das der Kapflora, das Waldgebiet und die Karroo, pflanzengeographisch vorgestellt. – In CHUN, C., (Ed.): Wissenschaftliche Ergebnisse der Deutschen Tiefsee- Expedition auf dem Dampfer "Valdivia" 1898–1899, **2(3)**, pp. 1–435. – Jena: Fischer.

MASSON, F., 1796: Stapelieae Novae: or a collection of several new species of that genus; discovered in the interior parts of Africa. (41 plates with text). – London: Nicol.

MEVE, U., 1988: *Duvalia vestita* MEVE (*Asclepiadaceae/Stapelieae*) – Eine neue Art aus der südwestlichen Kapprovinz. – Kakteen Sukk. **39:** 194–197.

– 1994: The genus *Piaranthus*. – Bradleya **12**: 57–102.

– 1995: A review of phytophagous insects on stapeliads (*Asclepiadaceae*). – Cimbebasia **14:** 103–106.

– ALBERS, F., 1990a: Die Stipularrudimente der *Stapelieae* (*Asclepiadaceae*). – Beitr. Biol. Pflanzen **65:** 99–107.

– 1990b: The species concept in *Duvalia* (*Asclepiadaceae*) – a preliminary revision of the genus. – Proceedings of the 12th meeting of the Association for the taxonomic study of the Flora of Tropical Africa, AETFAT. – Mitt. Inst. Allg. Bot. Hamburg **23b:** 595–604.

– LIEDE, S., 1994: Floral biology and pollination in stapeliads – new results and a literature review. – Pl. Syst. Evol. **192:** 99–116.

– ALBERS, F., KUSCH, G., 1990: The outer epidermal wall structure of African *Stapelieae* (*Asclepiadaceae*). – Nordic J. Bot. **9:** 519–523.

NEWTON, L. E., 1980: Phytogeographical associations of the succulent plant flora of south-west Arabia and the Horn of Africa. – Natl. Cact. Succ. J. **35:** 83–88.

PHILLIPS, E. P., 1932: *Duvalia maculata*. – Fl. Pl. Africa **12:** t. 474.

PLOWES, D. C. H., 1982: The stapeliads of South West Africa/Namibia. – Excelsa Taxon. Ser. **10:** 5–26.

RAHMAN, M. A., WILCOCK, C. C., 1991a: A micromorphological study on seeds of some Bangladesh asclepiads. – Bangladesh J. Bot. **20:** 55–60.

– 1991b: A report on flavonoid investigation in some Bangladesh asclepiads. – Bangladesh J. Bot. **20:** 175–178.

RAVEN, P. H., AXELROD, D. I., 1974: Angiosperm biogeography and past continental movements. – Ann. Missouri Bot. Gard. **61:** 539–673.

REESE, G., 1971: Untersuchungen über die Chromosomenzahlen der *Stapelieae* II. – Portugaliae Acta Biol., Sér. A **12:** 1–23.

– KRESSEL, H., 1967: Untersuchungen über die Chromosomenzahlen der *Stapelieae*. – Portugaliae Acta Biol., Sér. A **10:** 33–54.

ROWLEY, G. D., 1984: Jacquin and the *Stapelieae*. – Bradleya **2:** 1–8.

SCHILL, R., JÄKEL, U., 1978: Beitrag zur Kenntnis der Asclepiadaceen-Pollinarien. – Trop. Subtrop. Pflanzenwelt **22:** 53–170.

SCHLECHTER, R., 1885: Beiträge zur Kenntnis südafrikanischer Asclepiadeen. – Bot. Jahrb. Syst. **20** (Beibl. 51): 1–56.

SCHULTES, J. A., 1820: Systema Vegetabilium **4:** 1–129. – Stuttgart: Cotta.

SCHUMANN, K., 1895: *Asclepiadaceae*. – In ENGLER, A., PRANTL, K., (Eds): Die Natürlichen Pflanzenfamilien, **4,** pp. 189–306. – Leipzig: Engelmann.

SCHWARZACHER, T., AMBROS, P., SCHWEIZER, D., 1980: Application of Giemsa banding to orchid karyotype analysis. – Pl. Syst. Evol. **134:** 293–297.

SIMS, J., 1803: *Stapelia radiata.* – Bot. Mag., t. 619. – London.

– 1809: *Stapelia elegans.* – Bot. Mag., t. 1184. – London.

SNOW, R., 1963: Alcoholic hydrochloric acid-carmine as stain for chromosomes in squash preparations. – Stain Technol. **38:** 9–13.

STACE, C. A., 1989: Plant taxonomy and biosystematics, 2nd edn. – London: Arnold.

STEBBINS, G. L., 1980: Evolutionsprozesse. – Stuttgart: Fischer.

STRAKA, H., 1979: Über die Spaltöffnungen der *Stapelieae.* – Kakteen Sukk. **30:** 246–247.

STUESSY, T. F., 1990: Plant taxonomy. – New York: Columbia University Press.

SUBRAMANIAN, S. S., NAIR, A. G. R., 1968: Flavonoids of some asclepiadaceous plants. – Phytochemistry **7:** 1703–1704.

SWARUPANANDAN, K., 1984: Taxonomic and ethnobotanical studies on *Asclepiadaceae* R. BR. of Kerala. – Ph. D. Thesis, Kerala Forest Research Institute, Kerala, India.

SYLLA, T., 1986: Anatomisch-morphologische Untersuchungen der Frucht- und Samenentwicklung an ausgewählten Asclepiadaceen. – Staatsexamensarbeit, University of Münster.

– ALBERS, F., 1989: Samenentwicklung und Samenmorphologie krautiger und sukkulenter *Asclepiadaceae.* – Bot. Jahrb. Syst. **110:** 479–492.

THUNBERG, C. P., 1794: Prodromus plantarum capensium **1.** – Uppsala.

TJIO, J. H., LEVAN, A., 1950: The use of oxyquinoline in chromosome analysis. – Anales Estac. Exp. Aula Dei **2:** 21–64.

TROLL, W., 1959: Neue Beiträge zur Kenntniss der Blütenstände und Blüten von *Ceropegia*–Arten. – Akad. Wiss. Mainz, Abh. Math. Naturwiss. Kl. **5:** 227–263.

– 1964: Die Infloreszenzen, **1.** – Stuttgart: Fischer.

VAN ZINDEREN BAKKER, E. M., Sr., 1978: Quaternary vegetation changes in southern Africa. – In WERGER, M. J. A., (Ed.): Biogeography and ecology of southern Africa, pp. 133–143. – The Hague: Junk.

VOLK, O. H., 1944/1949: Zur Kenntnis der Pollinien der Asclepiadaceen. – Ber. Deutsch. Bot. Ges. **62:** 68–72.

– 1951: Zur Kenntnis der Stapelieen-Blüten. – Sukkulentenkunde **1** (1947): 46–59.

WALTER, H., 1973: Allgemeine Geobotanik. – Stuttgart: Ulmer.

WERGER, M. J. A., 1978: Biogeographical division of southern Africa. – In WERGER, M. J. A., (Ed.): Biogeography and ecology of Southern Africa, pp. 147–170. – The Hague: Junk.

WERTEL, H. P., 1976: Vergleichend-morphologische und entwicklungsgeschichtliche Untersuchungen über den Bau und Stellung der Infloreszenzen bei einigen Stapelieen-Gattungen. – Trop. Subtrop. Pflanzenwelt **17:** 264–330.

WHITE, A. C., SLOANE, B. L., 1937: The *Stapelieae* **1–3**, 2nd edn.– Pasadena: Abbey Press.

– 1942: An oddity in botanical nomenclature. – Cact. Succ. J. (Los Angeles) **14:** 158–159.

WHITE, F., 1983: The vegetation of Africa. – Paris: UNESCO.

ZEMKE, E., 1939: Anatomische Untersuchungen an Pflanzen der Namibwüste (Deutsch–Südwestafrika). – Flora **133:** 365–416.

ZOHARY, M., 1973: Geobotanical foundations of the Middle East **1** & **2.** – Stuttgart: Fischer.

Index to descriptions of taxa accepted and synonyms

(Taxa accepted in this paper are printed in bold face).

Caralluma anemoniflora = *Stapelia anemoniflora* 125
Duvalia anemoniflora = *Stapelia anemoniflora* 124
D. angustiloba 49
D. caespitosa 52
D. caespitosa var. ***caespitosa*** 52
D. caespitosa var. ***compacta*** 59
D. caespitosa var. *hirtella* = ***D. caespitosa*** var. ***caespitosa*** 52
D. compacta = ***D. caespitosa*** var. ***compacta*** 59
D. concolor = (?) ***D. caespitosa*** var. ***caespitosa*** 125
D. corderoyi 65
D. dentata = ***D. polita*** 92
D. eilensis 108
D. elegans 70
D. elegans var. *namaquana* = ***D. pubescens*** 99
D. elegans var. *seminuda* = ***D. elegans*** 70
D. elegans var. *seminuda* f. *magnicorona* = ***D. elegans*** 70
D. emiliana = ***D. caespitosa*** var. ***caespitosa*** 53
D. galgallensis 110
D. glomerata = (?) ***D. caespitosa*** var. ***caespitosa*** 125
D. gracilis 75
D. hirtella = ***D. caespitosa*** var. ***caespitosa*** 52
D. hirtella var. *minor* = ***D. caespitosa*** var. ***caespitosa*** 52
D. hirtella var. *obscura* = ***D. caespitosa*** var. ***caespitosa*** 52
D. immaculata 78
D. jacquini = ***D. elegans*** 70
D. jacquiniana = ***D. elegans*** 70
D. laevigata = (?) ***D. caespitosa*** var. ***caespitosa*** 125
D. maculata 81
D. maculata var. *immaculata* = ***D. immaculata*** 78
D. marlothii = ***D. caespitosa*** var. ***caespitosa*** 53
D. mastodes = ***D. caespitosa*** var. ***compacta*** 59
D. mastodis = ***D. caespitosa*** var. ***compacta*** 59
D. minuta = ***D. maculata*** 81
D. modesta 84
D. mustodes = ***D. caespitosa*** var. ***compacta*** 59
D. parviflora 87
D. pillansii 90
D. pillansii var. *albanica* = ***D. pillansii*** 90
D. polita 92
D. polita var. *parviflora* = ***D. polita*** 92

D. polita var. *polita* f. *intermedia* = ***D. polita*** 92
D. polita var. *transvaalensis* = ***D. polita*** 92
D. propinqua = ***D. caespitosa*** var. ***caespitosa*** 53
D. pubescens 99
D. pubescens var. *major* = ***D. pubescens*** 99
D. radiata = ***D. caespitosa*** var. ***caespitosa*** 52
D. radiata var. *hirtella* = ***D. caespitosa*** var. ***caespitosa*** 52
D radiata var. *minor* = ***D. caespitosa*** var. ***caespitosa*** 52
D. radiata var. *obscura* = ***D. caespitosa*** var. ***caespitosa*** 52
D. reclinata = ***D. caespitosa*** var. ***caespitosa*** 52
D. reclinata var. *angulata* = ***D. caespitosa*** var. ***caespitosa*** 52
D. reclinata var. *bifida* = ***D. caespitosa*** var. ***caespitosa*** 52
D. replicata = ***D. caespitosa*** var. ***caespitosa*** 52
D. somalensis = ***D. sulcata*** subsp. ***somalensis*** 118
D. sulcata 113
D. sulcata subsp. ***seminuda*** 116
D. sulcata subsp. ***somalensis*** 118
D. sulcata subsp. ***sulcata*** 113
D. sulcata var. *seminuda* = ***D. sulcata*** subsp. ***seminuda*** 116
D. transvaalensis = ***D. polita*** 92
D. transvaalensis var. *parviflora* = ***D. polita*** 92
D. tuberculata = (?) ***D. caespitosa*** var. ***caespitosa*** 125
D. velutina 121
D. vestita 102
Stapelia anemoniflora = *Duvalia anemoniflora* 124
Stapelia barbata = (?) ***D. caespitosa*** var. ***caespitosa*** 125
St. caespitosa = ***D. caespitosa*** var. ***caespitosa*** 52
St. concolor = (?) ***D. caespitosa*** var. ***caespitosa*** 125
St. corderoyi = ***Duvalia corderyoi*** 65
St. cymosa = (?) ***D. caespitosa*** var. ***caespitosa*** 125
St. elegans = ***Duvalia elegans*** 70
St. glomerata = (?) ***D. caespitosa*** var. ***caespitosa*** 125
St. hirtella = ***D. caespitosa*** var. ***caespitosa*** 52
St. jaquiniana = ***D. elegans*** 70
St. laevigata = (?) ***D. caespitosa*** var. ***caespitosa*** 125
St. radiata JACQ. = ***D. elegans*** 70
St. radiata SIMS = ***D. caespitosa*** var. ***caespitosa*** 52
St. reclinata = ***D. caespitosa*** var. ***caespitosa*** 52
St. replicata = ***D. caespitosa*** var. ***caespitosa*** 52

GPSR Compliance

The European Union's (EU) General Product Safety Regulation (GPSR) is a set of rules that requires consumer products to be safe and our obligations to ensure this.

If you have any concerns about our products, you can contact us on ProductSafety@springernature.com

In case Publisher is established outside the EU, the EU authorized representative is:

Springer Nature Customer Service Center GmbH
Europaplatz 3
69115 Heidelberg, Germany

Batch number: 10578403

Printed by Printforce, the Netherlands